# The Sputniks Crisis and Early United States Space Policy

## A Critique of the Historiography of Space

Rip Bulkeley

MACMILLAN

First published 1991

Published by
MACMILLAN ACADEMIC AND PROFESSIONAL LTD
Houndmills, Basingstoke, Hampshire RG21 2XS
and London
Companies and representatives
throughout the world

British Library Cataloguing in Publication Data
Bulkeley, Rip
The Sputniks crisis and early United States Space policy:
a critique of the Historiography of space. – (Studies in
Military and Strategic History)
1. Outer space. Exploration by United States, History
I. Title. II. Series
919.904
ISBN 978-1-349-11983-7     ISBN 978-1-349-11981-3 (eBook)
DOI 10.1007/978-1-349-11981-3

STUDIES IN MILITARY AND STRATEGIC HISTORY

General Editor: Michael Dockrill, Senior Lecturer in War Studies, King's College, London.

This major series of books on nineteenth and twentieth-century military, naval and air history explores hitherto neglected areas of the subject or provides fresh interpretations of existing material. The series covers the whole range of issues – strategic, diplomatic, economic and financial – involved in preparations for the conduct of, and the ending of, wars.

*Published Titles*

R. J. Q. Adams (*editor*)
THE GREAT WAR, 1914–1918: Essays on the Military, Political and Social History of the First World War

Rip Bulkeley
THE SPUTNIKS CRISIS AND EARLY UNITED STATES SPACE POLICY:
A Critique of the Historiography of Space

David A. Charters
THE BRITISH ARMY AND JEWISH INSURGENCY IN PALESTINE, 1945–1947

Andrew J. Crozier
APPEASEMENT AND GERMANY'S LAST BID FOR COLONIES

David Devereux
BRITISH DEFENCE POLICY TOWARDS THE MIDDLE EAST, 1948–1956

John Robert Ferris
THE EVOLUTION OF BRITISH STRATEGIC POLICY, 1919–1926

Alfred Gollin
THE IMPACT OF AIR POWER ON THE BRITISH PEOPLE AND THEIR GOVERNMENT, 1909–1914

This book is dedicated to all who have helped me to learn the crafts of writing and research, and especially to S. H. Burton, Lawrence Freedman, Christopher Meredith, Paul Rogers and E. P. Thompson, and to the memory of John Pierson Bulkeley.

# Contents

# Preface

During the 'Star Wars' debates of the 1980s I became increasingly curious about the underlying attitudes towards space policy of people in the United States. The more I learned about the field the more likely it seemed that the irrational but indefeasible anxieties expressed in such books as *Soviet Conquest from Space* by Peter James, or *Strike from Space* by Phyllis Schafly and Chester Ward, had at least part of their origins in the trauma of the American people's first frustrating and humiliating encounter with the problems of seeking political advantage from would-be technological feats in space, the sputniks crisis of 1957–8.[1]

According to Walt Rostow:

> There is no clear analogy in American history to the crisis triggered by the launching of the Soviet earth satellite on October 4, 1957. This intrinsically harmless act of science and engineering was also, of course, both a demonstration of foreseeable Soviet capability to launch an ICBM and a powerful act of psychological warfare. It immediately set in motion forces in American political life which radically reversed the Nation's ruling conception of its military problem, of the appropriate level of the budget, and of the role of science in its affairs. The reaction reached even deeper, opening a fundamental reconsideration not only of the organization of the Department of Defense but also of the values and content of the American educational system and of the balance of values and objectives in contemporary American society as a whole.[2]

Taking Rostow's sketch of the sputniks crisis as a list of possible topics for a book with the present title, it is only fair to warn the potential reader that several will not be covered. The sequence of events discussed in these pages does not extend far beyond the early months of 1958. Nothing will be found here about the origins of NASA or the 1958 National Defense Education Act, for example, and very little about any aspect of the Eisenhower administration's positive space policy from 1958 to 1960. It has even been necessary to neglect certain aspects of the immediate sputniks crisis, such as President Eisenhower's failing health and the internal politics of the Democratic Party. On the other hand four chapters are given to the embryonic space policy of the United States under Eisenhower's predecessor, Harry Truman. The reason for this is that the focus of my interest has been the enduring influence of the sputniks crisis on American perceptions of the earliest phase of their country's space effort. As a powerful act of psychological warfare the sputniks should have left some mark on American culture, and the most likely place to find it was in the writing of space history itself, a discipline in which the bulk of original scholarship to date has been contributed by Americans.

From an early stage in my research I was struck by the obscurity of the factual record of the development of American space policy between 1945 and 1957. Where possible I have done what I can to sort things out. This has prevented me from conforming with the convention under which scholars do not correct each other's mistakes in public, which is to say, in the presence of the fellow-citizens who depend on their expertise and who frequently pay their salaries. There was also the important consideration that widespread carelessness with the facts of the pre-sputnik period might itself be evidence for the persistence of an overly subjective historiographical tradition. It is surely right to be concerned for the intellectual health of a scholarly discipline within which a great American university can publish a historical study, apparently endorsed by two of the most eminent historians of post-war American science, which ignores both the vertical V-2 flights of the 1940s and the first two sputniks, to proclaim without even an argument that *American* scientists and engineers launched the first vertical sounding rocket to enter space and even, at one point, the world's first artificial satellite.[3]

But I must also admit that my own attempts at resolving factual difficulties have had only limited success, for several reasons. Though American archives have a well-earned reputation for their openness and generosity to visiting researchers, many important documents remain classified in whole or in part. A foreigner working in this field faces the additional obstacle that some important secondary sources, to which most American scholars have access, are still restricted. In my case there were also severe financial constraints on the amount of time I could give to primary research in the United States, and hence on the number of interviews I could conduct and the number of archives I could visit. This is of course a problem for all historical researchers to some extent. But it was exacerbated in this instance by the degree of factual unreliability in the secondary literature over matters ranging from minor details, such as the date of an international scientific meeting, to key events such as an official Soviet statement. Thus after doing the best I could with my·limited skills and resources I am aware that many errors must remain, to which I would welcome corrections, provided of course that those who tender them shall have first consulted the evidence cited here, and not merely the oral tradition of what too many people now 'know' to have happened between 1945 and 1957, but which frequently did not. (Having departed so often from the conventional wisdom, I have taken care wherever possible to provide references to primary sources in sufficient detail for them to be checked with ease.)

To fellow peace researchers and campaigners who may be discomfited by my tendency, for the sake of narrative convenience, to assume the strategic values and global perspective of bygone governments of the United States and their advisers, I can only plead that I have felt it more important to deliver some actual grounds for the revision of a historical tradition which still suffers from its origins in the Cold War, than to preach against the familiar lunacies of the nuclear arms race. I hope I have done enough to show that, while it may yet be too early to rescue space history entirely from the corruption of cold-war attitudes, it is not too early to try.

RIP BULKELEY
*April 1990*

# Acknowledgements

Research for this book was conducted in. the following libraries and archives: *in Britain* – Bodleian Law Library; Bodleian Library; British Library, Bloomsbury; British Library of Economic and Political Science; British Interplanetary Society Library; British Newspaper Library; International Institute of Strategic Studies; Radcliffe Science Library; Rhodes House Library; Royal Society Library; Science Museum Library; Science Reference Library; US Information Service London Reference Center; University College Library; and the World Data Centre C1 (Geophysical and Solar), Rutherford-Appleton Laboratory. *In the United States* – Boston Public Library; Dwight D. Eisenhower Library; Georgetown University Library; Harry S. Truman Library; Library of Congress; Lyndon Baines Johnson Library; MIT Library; NASA History Office; National Academy of Sciences; National Air and Space Museum Library; and National Archives. I am grateful for the help of staff at all these institutions, and particularly for that of Dr Lee Saegesser at the NASA History Office and Dr David Haight at the Eisenhower Library.

I would like to thank the Cincinnati Historical Society for permission to reproduce from its *Bulletin* an excerpt from the diary of Oliver Gale (p. 190), the NASA History Office for permission to quote from transcripts of Oral History interviews with E. M. Galloway, B. N. Harlow, H. E. Newell, R. M. Salter, and G. W. Siegel, and the Columbia University Oral History Research Office for permission to cite 'The Reminiscences of Richard Bissell, 1967', the transcript of an interview in the Oral History Collection of Columbia University. Passages from transcripts of Oral History interviews conducted with Professor James Van Allen and Dr Charles Johnson as part of the Space Astronomy Oral History Project at the National Air and Space Museum, Washington, have been reproduced with the general consent of the Project. Since copyright rests with interviewees in the latter case every effort was made to secure their explicit permission and the author is grateful to Professor Van Allen for having granted it. Unfortunately it has not been possible to contact Dr Johnson directly, but since his interview was assigned 'public' status within the Project and the quotation has been made in good faith it is assumed that no objection will be raised.

I am especially grateful to all those who agreed to be interviewed as part of my research (p. 216), and to the following correspondents for their help with particular details or general background: Dr Ray Cline, Dr Desmond King-Hele, Sir Bernard Lovell, Dr Allan Needell, Professor Alan Shapley, and Dr Hans Ziegler.

I would also like to express my gratitude to my wife, Jane Bulkeley, for her constant support, to the Institute of Policy Studies for its generous hospitality in Washington, and to Fran Bagenal, Alex Godden, Ellen Gruenbaum, Gerry Hale, Hayden Peake, and Paul Stares for their assistance with various matters. During the initial gestation of this book as a doctoral thesis written at King's College London I received much needed encouragement and good advice from Professor Lawrence Freedman. I hope that some of it at least has been put to good use.

Needless to say none of the individuals or institutions named above is in any way responsible for my output from their input.

RIP BULKELEY

# List of Abbreviations
# and Technical Terms

Familiar abbreviations and acronyms such as 'Ltr', 'J.', 'CIA', or 'USAF' and any acronym used only in a single passage and explained at that point have been omitted. A separate list of abbreviations and acronyms for archival sources is given in the Notes (p. 215).

| | |
|---|---|
| AAF | US Army Air Forces. |
| AEC | US Atomic Energy Commission. |
| aerodynamics | study of the motion and control of solid bodies in air, and of air in motion. |
| aeronomy | study of the physics and chemistry of the upper atmosphere. |
| AFB | US Air Force Base. |
| AMC | USAF Air Materiel Command. |
| *Annals* | *Annals of the IGY.* |
| APL | Applied Physics Laboratory, Johns Hopkins University. |
| apogee | maximum altitude of missile trajectory or satellite orbit. |
| ARDC | USAF Air Research and Development Command. |
| ARS | American Rocket Society. |
| | |
| Biog. | Biographical. |
| BoB | White House Bureau of the Budget. |
| BuAer | US Navy Bureau of Aeronautics. |
| | |
| CalTech | California Institute of Technology. |
| CSAGI | Comité Spécial de l'Année Géophysique Internationale. |
| | |
| DCI | Director of Central Intelligence. |
| DEW Line | Distant Early Warning (radar) Line. |
| DOD | US Department of Defense. |
| DOS | US Department of State. |
| DTM | Department of Terrestrial Magnetism, Carnegie Institution. |

| | |
|---|---|
| GPO | US Government Printing Office, Washington. |
| | |
| HMSO | His Majesty's Stationery Office. |
| House | US Congress, House of Representatives. |
| | |
| IAF | International Astronautical Federation. |
| IATME | International Association of Terrestrial Magnetism and Electricity. |
| IAU | International Astronomical Union. |
| ICBM | Inter-Continental Ballistic Missile. |
| ICIC | Interdepartmental Committee on Interplanetary Communications, Soviet Academy of Sciences. |
| ICSU | International Council of Scientific Unions. |
| IGY | International Geophysical Year. |
| IPY2/3 | Second/Third International Polar Year. |
| IRBM | Intermediate-Range Ballistic Missile. |
| IRE | Institute of Radio Engineers. |
| IUGG | International Union of Geodesy and Geophysics. |
| | |
| JIC | Joint Intelligence Committee. |
| JPL | CalTech Jet Propulsion Laboratory. |
| JRDB | US War and Navy Departments, Joint Research and Development Board. |
| | |
| mantle | region of Earth's interior between the outer lithosphere or crust and the central molten core. |
| mass ratio | ratio of the mass of rocket propellent to the total mass of the rocket including propellent. |
| MCI | ICSU Mixed Commission on the Ionosphere. |
| Mc/s | megacycles, measure of frequency at one million cycles per second (= megahertz/MHz). |
| MIT | Massachusetts Institute of Technology. |
| MRBM | Medium-Range Ballistic Missile. |
| | |
| NACA | National Advisory Committee on Aeronautics. |
| NAS | US National Academy of Sciences. |
| NASA | National Aeronautical and Space Administration. |
| NRL | Naval Research Laboratory. |
| NSC | National Security Council. |
| NSF | National Science Foundation. |

| | |
|---|---|
| OASD | Office of the Assistant Secretary of Defense. |
| OCB | Operations Coordinating Board. |
| ODM | White House Office of Defense Mobilization. |
| ODM-SAC | ODM Science Advisory Committee. |
| ONR | Office of Naval Research. |
| OSD | Office of the Secretary of Defense. |
| OSRD | Office of Scientific Research and Development. |
| | |
| perigee | lowest point of satellite orbit. |
| PSAC | US President's Science Advisory Committee. |
| | |
| R&D | Research and Development. |
| radio interferometer | radio receiver with an array of antennas designed to provide information about the structure of a large radio source, or the trajectory of a mobile point-source such as a satellite-borne transmitter, by comparing the phases (or times) at which its signal is received on different antennas. |
| RAFVR | Royal Air Force Volunteer Reserve. |
| RCA | Radio Corporation of America. |
| RDB | US Department of Defense, Research and Development Board. |
| | |
| SAB | AAF/USAF Scientific Advisory Board. |
| SACEUR | Supreme Allied Commander, Europe. |
| SHAPE | Supreme Headquarters, Allied Powers in Europe. |
| SLV | Satellite Launch Vehicle. |
| SMEC | Strategic Missiles Evaluation Committee. |
| specific impulse | ratio between thrust produced and rate of consumption of rocket propellent. |
| | |
| TCP | ODM-SAC Technological Capabilities Panel. |
| TEG | RDB Committee on Guided Missiles, Technical Evaluation Group. |
| telemetry | technique of measuring a physical quantity and transmitting the results automatically to a distant receiving station. |
| TPESP | US IGY Committee, Technical Panel on the Earth Satellite Program. |
| TV | Test Vehicle. |

| UFO | Unidentified Flying Object. |
| UNESCO | United Nations Educational, Scientific, and Cultural Organization. |
| URSI | International Scientific Radio Union. |
| USIA | US Information Agency. |
| USNR | US Naval Reserve. |
| WDC | IGY World Data Centre. |
| WMO | World Meteorological Organization. |

You could write a whole book about 1945 to 1957.

Eugene M. Emme, first NASA Historian, 1959–78

In the long run, it is impossible to stand in the way of the exploration of truth. Someone will learn, somewhere, sometime.

Philip Handler, president, US National Academy
of Sciences, 1969–80

# Part I Crisis

Part I  Labels

# 1 A Devastating Blow

## 1 THE IMPACT OF THE SPUTNIKS

The world's first artificial satellite, Sputnik 1, was launched at four minutes to midnight (local time) on 4 October 1957 from a recently completed rocket testing facility in the desert east of the Aral Sea, near the town of Tyuratam in the Kazakh Soviet Socialist Republic.[1] In Moscow the time was 10.26 p.m., in London 7.26 p.m. In the eastern time zone of the United States it was the middle of the afternoon on the last but one day of an international scientific conference which had assembled to discuss the uses of rockets and artificial satellites within a programme for investigating physical properties and processes of the Earth as a complete planet, the International Geophysical Year or IGY, which had been running since 1 July 1957.

Sputnik 1 carried a powerful radio beacon and twice passed within easy detection range of receiving stations in the United States before anyone in that country knew of its existence. Radio Moscow first broke the news almost three hours after the launch, at 1.22 a.m. Moscow time. Eighteen minutes later a detailed report from the TASS news agency was read over the air. The BBC monitoring service at Caversham, Reuters news agency, and the Moscow bureau of *The New York Times* led the scramble to break the news in the West. At about 6.15 p.m. Eastern Daylight Time (EDT) it was telephoned through to Walter Sullivan, the senior science correspondent of *The New York Times*, who was attending a reception for delegates to the IGY Rockets and Satellites Conference at the Soviet Embassy in Washington. Sullivan took his sensational information to fellow-guests William Pickering, director of the Jet Propulsion Laboratory, and Lloyd Berkner, the American Vice-President of the IGY's international organizing committee and Reporter (coordinator) for rockets and satellites work within the Year. Berkner then announced the Soviet success to the assembled scientists and officials and congratulated his Soviet colleagues on their momentous achievement.

At 8.07 p.m. EDT the powerful receivers of the Radio Corporation of America at Riverhead, Long Island, became the first installation in the United States to pick up radio signals from Sputnik 1, by then on its fourth pass over the Western Hemisphere. During the night that followed

3

thousands of amateur optical observers in 'Moonwatch' teams all over North America were alerted by their coordinators at the Smithsonian Astrophysical Observatory in Cambridge, Massachusetts. But satellites in low orbit can only be seen around dusk and dawn, and the orbit of Sputnik 1 was such that, at first, neither the satellite proper nor the much larger upper stage of its carrier rocket, which was also placed in orbit, could be spotted visually from the United States. The first sighting in the United States, probably of the rocket upper stage, was by a British astronomer, Gordon Little, at the University of Alaska's Geophysical Institute at 5.01 a.m. local time (11.01 EDT) on Sunday, 6 October.

At its closing session on 5 October the IGY conference heard further details of the Soviet satellite project from their chief delegate, Anatoli Blagonravov, who had already given a presentation on it two days earlier. The meeting naturally extended its formal congratulations to the Soviet Academy, and this sentiment was echoed a few days later in a statement issued by President Eisenhower. Even the most objective and dispassionate American scientists, however, had mixed feelings about the news of Sputnik 1. James Van Allen, a distinguished atmospheric physicist who was at sea on a US Navy ship seconded to an IGY rocket-launching programme when the news came through, made a careful note of his immediate thoughts. After acknowledging the 'brilliant achievement' of his Soviet colleagues his first thought was that the satellite was a tremendous, and by implication regrettable, propaganda coup for the Soviet Union. This was followed by recognition of the significance of the far greater weight to orbit which had been achieved by the Soviet system, as compared with its still untried American counterpart. Next came retrospective disgust because, in his opinion, the United States' humiliation had been quite unnecessary.[2] Van Allen clearly shared the view of many of his contemporaries that Soviet victory in the first phase of the 'space race' was largely due to the fact that at the outset of the programme in 1955 a panel of American scientists had recommended an untried and over-elaborate rocket design for the American satellite project, in a split decision where the vote went against a minority which supported a slightly less ambitious system using more or less off-the-shelf components, albeit in a new combination.

In a more positive spirit than Van Allen's, one unnamed American scientist remarked immediately after the news that 'Maybe the Russian-American competition hasn't been so bad on the satellite if it encourages us into beating them to the moon.'[3] Shrewd as this comment may have been, it was merely voicing a thought which had already been circulating amongst American scientists interested in space exploration for about five years.[4]

American politicians, as is their wont, were rather more blunt in responding to Sputnik 1 as an unwelcome, almost an unfair surprise, particularly since it confirmed the Soviet claim, issued a few weeks earlier, to have successfully tested the world's first intercontinental ballistic missile (ICBM), which some opinion-makers had tried to pooh-pooh at the time. Senator Henry 'Scoop' Jackson, the Democratic chairman of the Military Applications Subcommittee of the Joint Committee on Atomic Energy, described the Soviet triumph as 'a devastating blow to the prestige of the United States as the leader of the scientific and technical world'.[5] Senator Stuart Symington, a Democrat on the Senate Armed Services Committee, saw the satellite as 'one more proof of growing Communist superiority in the all-important missile field. If this now known superiority develops into supremacy, the position of the free world will be critical.'[6] Both he and Jackson placed the blame at the door of Eisenhower's Republican administration, which in their opinion had cut spending on missile programmes too severely. By contrast, Senator Styles Bridges, the senior Republican on the Senate Armed Services Committee, preferred to draw attention to the influence on defence budgets of short-sighted public attitudes:

The time has clearly come to be less concerned with the depth of the pile on the new broadloom rug or the height of the tail fin on the new car and be more prepared to shed blood, sweat and tears if this country and the free world are to survive.[7]

In general, American citizens were not immediately alarmed by the news of Sputnik, but public concern grew rapidly over the weekend in response to the torrent of anxious commentary, much of it hostile to the record of the Eisenhower administration, which flowed from every part of the news media. Thus the statement by a reporter for the London *Sunday Express*, that Sputnik 1 had thrown the United States 'into frantic, fearful and angry confusion', may not have been entirely accurate when it was published on 6 October, but it became rapidly more so as Americans were persuaded to perceive the event as a national crisis. Vice-President Richard Nixon, for example, declared that Sputnik 1 was 'a grim and timely reminder . . . that the Soviet Union has developed a scientific and industrial capacity of great magnitude', and the industrial magnate and elder statesman Bernard Baruch announced that 'Suddenly, rudely, we are awakened to the fact that the Russians have outdistanced us in a race we thought we were winning.'[8] The movement of public opinion was strongly reinforced by Soviet statements to the effect that a second larger satellite would shortly be launched, and then by the actual launching of Sputnik 2 on 3 November 1957, carrying a live dog which survived for seven days in space.

People demanded to know what avoidable mistakes had been made in America's post-war research and development (R&D) programmes, and by whom. And, as the examples already quoted show, the political potential of the sputniks issue, both positive and negative, was immediately apparent to many people, above all to the Democrats. George Reedy later wrote that 'After five years of a Republican administration there was little doubt as to the doorstep upon which the blame would be laid.'[9] He ought to know, since as director of the Senate Democratic Policy Committee and a close personal adviser to presidential aspirant Senator Lyndon Johnson he had much to do with making things turn out that way at the time.[10]

The post-sputnik investigative and explanatory reporting of the American media was often, if not always, of a high standard, and their work, together with the comments of public figures to which they gave an airing, laid the foundations for Congressional hearings held in the winter and spring of 1957–8, prominent amongst which were those mounted by the Preparedness Investigating Subcommittee of the Senate Armed Services Committee, chaired by Senator Johnson.[11] Gradually, during 1958, the post-mortem on the sputniks gave way to more positive political and legislative activity, culminating in the establishment of the National Aeronautical and Space Administration (NASA). But for the student of pre-sputnik space policy it is those first inquiries, both formal and informal, which hold the greater interest.

In the immediate furore after 4 October 1957 the press found many well-informed witnesses willing to accuse the administration of negligence, including some of its own former officials. As Assistant Secretary for Air for R&D from 1953 to 1956, Trevor Gardner had galvanized the American missile programme into a major effort during Eisenhower's first term. On 16 October, however, he told the New York *Herald Tribune* that the satellite project had been seriously underfunded and had received a priority 'inconsistent with the international gamble for our technological reputation'. When asked whether 'some of the blame for the United States' lag goes back as far as the Truman administration?' he answered with a flat negative.[12] His candour did not, however, extend to reminding the interviewer that while in office he himself had dismissed the satellite project as 'just another of the programs that interfere' with the development of ICBMs.[13]

A few days later *Life* published an article by Clifford Furnas, who had served as Assistant Secretary of Defense for R&D from December 1955 to February 1957, the crucial development years of the American IGY satellite project. Furnas ignored anything which had happened before '1954, [when] plans got under way for the International Geophysical Year'. He blamed America's failure to be first into space on the 'tragically naive and shortsighted outlook' of the Department of Defense, which had regarded basic research 'as a sort of extracurricular scientific pastime to be indulged in only if money is left over from the "really important" things'. In his opinion, that attitude had been personified in the recently retired Secretary of Defense, Charles Wilson.[14]

\*        \*        \*

There was, however, another and very different point of view, summed up by the comment in *Time* magazine that the failure of the United States to be first into space had been due to 'eight lost years after World War II', namely, under President Truman.[15] The most impressive exponent of this opinion was none other than Wernher von Braun, the former technical director of Hitler's war-time missiles project, who had been working for the US Army since 1945. He pointed out that 'the United States did relatively little about rocket research until the beginning of the Korean war'.[16] Truman was so offended by the imputation of neglect that he replied in person, declaring that 'as soon as [he] saw that program was lagging due to interservice rivalry' he had placed 'one of the top production engineering men of industry . . . [Chrysler's president Kaufman Keller] in full charge', with completely satisfactory results. He added that Keller had then been 'one of the first to be dismissed' by the incoming Eisenhower administration, with the result that interservice rivalry had once more reared its hydra heads.[17] But this account was rebutted in turn by Arthur Krock, who had been chief of the Washington bureau of *The New York Times* at the time. Krock pointed out that Keller had acted only as a consultant, although one whose advice was almost always accepted, and that he had resigned of his own volition after continuing to serve under Eisenhower for about eight months.[18]

In a follow-up piece, Krock quoted extensively from the two memoranda by John McCone, then Under Secretary for Air for R&D, which had led to Keller's appointment. However, Krock neglected to point out that the Truman administration's emphasis on short-range air-defence missiles had virtually mandated Keller's disregard for the ICBM.[19] As will be shown in Chapter 6, the Truman administration left its successor quite poorly equipped to begin launching satellites a few years after taking office in January 1953. The only ballistic missile under development at that point was the Redstone, the first new weapon which von Braun and his fellow-engineers at the Army's Redstone Arsenal in Huntsville, Alabama, had been allowed to develop. This was at best a marginal basis for a space-launch system.

Truman and von Braun maintained their disagreement with further statements in November and December.[20] Other politicians jumped into the fray. Paul Butler, chairman of the Democratic National Committee, said that he blamed the present Republican administration, but that since von Braun had been 'in charge of the whole program' under Truman (which had not been Truman's story) any responsibility dating back to that period lay with von Braun alone.[21] Congressman Gerald Ford, the senior Republican member of the House Subcommittee on Defense Appropriations, responded that the key decision which had prevented the United States from acquiring an ICBM by 1953 had been the Truman administration's cancellation of a contract for ICBM studies with Convair in June 1947, and its subsequent failure to reinstate that project (the MX-774) in any form until January 1951.[22] According to Ford, Eisenhower as Chief of Staff had advised against this at the time. This claim was supported with testimony given by Eisenhower in 1946 and 1947 which, while not explicitly referring to ICBMs, had certainly singled out 'the fields of guided missiles, electronic devices and high-speed high-performance aircraft' for particular attention.[23]

To refute Butler von Braun needed only to point out that he had not even been an American citizen at the time he was said to have held the nation's missile programme in his hands. At the same time, however, he began to soften the implications of his earlier statement that the lack of American ballistic missile programmes between 1945 and 1951 had been 'the main reason why the Soviet Union was first into Space'.[24] He now claimed that:

> . . . the United States could not have undertaken major military projects involving large expenditures during that period. The war was over and the people wanted homes, autos and other things, not long-range rockets.[25]

In fact, of course, even before the outbreak of the Korean War in June 1950 the United States had undertaken several 'major military projects involving large expenditures', from the development and production of the B-36 bomber to the Berlin Airlift. Even the relatively minor sums expended in the vain pursuit of a nuclear-powered bomber might have made a considerable difference to the outcome of the first space race, if they had been used to develop ballistic missiles instead.[26] But von Braun was probably less concerned with factual accuracy than with mending his fences with the increasingly 'pro-space' Democrats.

When he appeared before the Preparedness Subcommittee a few days later von Braun still maintained that the six years 'where we did not work on the Redstone missile' had been decisive. But he also repeated the excuse about the allegedly relaxed, peacetime atmosphere of the late 1940s. The Truman delay had been 'entirely understandable. We all thought the war was over.' While an ICBM *could* have been developed by 1950, circumstances had ruled out the necessary intensive effort.[27]

*        *        *

Not surprisingly the argument about which side of the American political system was to blame for the sputniks crisis ran and ran. Truman's defence of his administration's achievements in the missile field remained unshaken by further criticism in a public memorandum signed by Senator William Knowland, the senior Republican member of the Preparedness Subcommittee, and Representative Joseph Martin, in response to 'the incredible remarks made by you in a hotel lobby in New York yesterday'' Knowland and Martin stated that independent testimony to the Subcommittee had shown that 'a considerable part of the blame lies on the doorstep of your Administration'. They also claimed that Truman had failed to act on a recommendation from the Air Policy Commission in January 1948 'that research in this field [the intercontinental missile] be given the highest priority'.[28] (In point of fact, the Commission's advice on missiles was ambivalent to a degree, and came far from recommending a crash programme to develop the intercontinental *ballistic* missile as such – see Chapter 6.)

The Knowland/Martin memorandum contained some impressive and probably valid figures:

> More than 99 percent of the money spent on long range missiles by the United States has been spent by the Eisenhower Administration – less than one percent by your [Truman's] Administration. Expenditures for long range ballistic missile development will have increased 1000 times between the last Truman year and the next fiscal year.[29]

However, the Republicans were faced from the outset with the problem that their preferred way of explaining the sputniks setback was simply not being accepted by the electorate. Instead, public opinion followed the apparent consensus of the experts as presented by the information media. In December 1957 the Gallup Poll asked people 'Where, specifically, would you put the blame, if anywhere, for letting the Russians get ahead of us in developing rockets and missiles?' The most popular answer was 'Eisenhower Administration, the Republicans, our leaders in Washington' and the least popular was 'Former Administrations – Truman, Roosevelt Administrations'.[30]

To add to Republican frustration, the hearings of the Preparedness Subcommittee, which were a focus of national attention for several weeks, were being conducted on terms which made it impossible for them to get their information across effectively. Throughout its proceedings missiles received far more attention than satellites, and the future course of conduct for the United States was discussed at much greater length than the recent past. This was not merely an understandable choice of priorities in the circumstances. It was also convenient for various parties that a veil be drawn over the early record on United States space policy.

The Democrats were manifestly anxious to protect the record of the Truman administration in respect of the development and application of large rockets. The press might air the question of the degree of responsibility for the sputniks crisis to be assigned to the previous administration, for example by recalling and evaluating the work of Kaufman Keller in its later years, but this side of the matter was glossed over at the Subcommittee, steered as it was by a Democratic chairman and his hand-picked staff. Keller was never called to testify before it.

Then there were the scientists who had originally talked the Eisenhower administration into mounting a satellite project for the IGY, some of whom had chosen the Vanguard launching system. With careers to build in the newly burgeoning field of space exploration, they naturally preferred to avoid having to explain their previous advice in any detail. The Subcommittee in fact called none of them to give formal testimony, relying instead on John Hagen, the director of Project Vanguard, who had had nothing to do with its initiation or the selection of a launch system. One scientist who had been heavily involved with the original choice of a satellite launch system in 1955 was Homer Stewart of CalTech's Jet Propulsion Laboratory. But instead of calling Stewart as a witness the Subcommittee appointed him their senior scientific adviser. In that capacity he was well placed to help direct the investigation, instead of being called before it as he might otherwise have been.

The Subcommittee called only two civilian officials with responsibility for rocket R&D who had held office before 4 October 1957. One was William Holaday, who had been responsible for the missile programmes of the Department of Defense, but not for the IGY satellite project. The other was Donald Quarles, who had been involved with the initiation and selection of Project Vanguard in 1955, as Assistant Secretary of Defense for R&D, but who was promoted to Secretary of the Air Force immediately afterwards. His successor as Assistant Secretary, Clifford Furnas, who had been responsible for supervising the actual execution and, equally important, the financing of Project Vanguard, was never called. Nor was the former Secretary of Defense, Charles Wilson, who had retired a few days after the launching of Sputnik 1, and who was charged by many (including Furnas) with having crucially obstructed the satellite project. Nor was the former Assistant Secretary for Air for R&D, Trevor Gardner, who had resigned in February 1956 in protest against what he saw as a wasteful multiplication and diversification of missile projects through interservice rivalry, and who had included the satellite project under that heading. In short, the Subcommittee markedly refrained from anything like a thorough and objective review of the development and implementation of the policies of the Eisenhower administration on missiles and satellites.

In the circumstances only one Republican could have used the historical record to counter Johnson's brilliantly 'non-partisan' exploitation of the space issue, which will be analysed in more detail in Chapter 12. This Eisenhower firmly refused to do:

> As I returned Dr Wiesner was telling the President that much of the problem relating to the timing on missiles and the satellite came from our late start on them – rather than from delays in their conduct after they were initiated by this Administration. The President said he has avoided going into that matter because it becomes immediately political, and he just does not want to do that.[31]

## 2 THE AIMS OF THIS BOOK

This book presents a fresh examination of the history of early American space policy, in the light of official papers some of which have only recently been declassified. The findings shed new light on the immediate post-sputnik recriminations in the United States, described above. It will be shown that the Republican point about Truman's neglect of missile projects was considerably more relevant than was generally conceded at the time. It will also be shown that the Eisenhower administration was by no means so heedless of the political stakes involved in space exploration as has been claimed by its detractors, then and since. Even more controversially it will be argued that American space historians have long tended to give way to prejudice and hearsay when dealing with the early post-war period, although some of the relevant evidence, at least, has been available for some time. In the past that bias has fuelled hawkish attitudes in many of those in the United States and elsewhere who concern themselves with space policy, and its influence continues to be felt today.

Thus the book is just as much about aspirations and perceptions as about the factual record of early American space policy. It is also of necessity an investigation of only one side in the first space race. When it comes to the history of *Soviet* space policy for the same period, from the end of World War 2 to the launching of the first sputniks, the continued non-availability of Soviet government papers means that nothing like sufficient accuracy and detail can be achieved for a reliable and historically satisfactory account of the other side of the story.[32] What is presented here is, therefore, a historical investigation not of the early sputniks themselves but rather of the immediate Western response to them. It will be argued in Chapter 13 that that original response has continued to dominate Western perceptions of the past course of events in space, and thereby to shape public attitudes and national policies in the strategic competition between East and West. In the author's opinion this tradition is long overdue for critical re-examination.

*                    *                    *

American space history was born in the climate of alarm and recrimination which followed the launching of the first sputniks. Some idea of the problems created by such origins may be conveyed by a brief survey of the books which were published in 1958 with the aim of explaining their recent humiliation to Americans.

The first to appear, in January 1958, was a collection of recent newspaper articles and interviews supplied with linking commentary by Richard Witkin, a reporter for *The New York Times*. Part of Teller's evidence to the Preparedness Subcommittee was included, but not his remarks about the regrettably late start in the American ICBM programme. Gardner's *Herald Tribune* interview, with its explicit exoneration of the Truman administration, was also selected. None of the other contributions referred to anything that had occurred before Eisenhower took office in January 1953.[33]

The books which followed Witkin's adopted various readings of the history of American rocket and space projects from 1945, but none would have caused any inconvenience for managers of the Democratic interest such as George Reedy. Lieutenant General James Gavin, a former head of the US Army's R&D command who resigned after Sputniks 1 and 2 over what he saw as the inadequacy of the American response, chose to ignore the Truman era completely.[34] So did the editors of the popular magazine *Missiles and Rockets*, despite the fact that their book included a preface by von Braun which repeated his charge that the main cause for the Soviet lead in space had been the American failure to start a ballistic missile development programme before 1951.[35] For *Missiles and Rockets* everything was the fault of the current incumbent of the White House, to whom they pontificated that: 'John Q. Public understands the grave and alarming mess we are in whether your Cabinet members do or not.'[36]

Others forsook the discretion of silence for the bold falsehood that the American R&D programme for military and other applications of rockets had been just as large and as vigorous as its Soviet equivalent ever since 1945, right through the Truman administration. This was asserted, for example, by Donald Cox and Michael Stoiko, in one of the first full-length studies of the political and military implications of space technology available to the general public.[37] Their view was echoed a few months later in an otherwise careful account of space technology compiled by Carsbie Adams of the National Research and Development Corporation, which stated that 'almost immediately after World War II the "cold war" began. The United States, like the Soviet Union, embarked on a *lavish* guided-missile research and development program.'[38]

However no one, not even authors who referred to the Truman period in such positive terms, ever mentioned the appointment of Kaufman Keller in October 1950, despite its obvious bearing on the issues of the day, and despite the fact that it had so recently been the subject of public exchanges between Truman and his Republican critics. Democrats and their supporters in the space-writing community were faced with a dilemma. First, as testimony to the effectiveness of Truman's handling of rocket programmes, Keller's appointment came rather late, since it implied that they had been handled poorly for the previous five years, or most of Truman's time in office. Second, if Keller had done as good a job as Truman claimed, the United States should have been quite well-placed in the race to develop ICBMs when Eisenhower took over. But that was manifestly not the case, and it was Eisenhower's administration, not Truman's, which had in fact taken the crucial decisions to start a crash programme to develop ICBMs, and which had spent vastly greater sums on missile projects than Truman ever did. In short, the actual details of Truman's handling of rocket research and development had to be avoided at all costs, if the denunciation of Eisenhower was to proceed effectively.

Within this general strategy of silence or vagueness various authors put forward more idiosyncratic explanations for the American lag behind

Soviet R&D on missiles during the Truman period. Cox and Stoiko claimed that 'Russia acquired almost all of Hitler's best rocket brains', a judgement which few later historians would ever support. They also adopted Walter Lippmann's explanation for the American failure to match the early Soviet space achievements, which had been that 'What McCarthyism did to the inner confidence of American scientists and thinkers has constituted one of the great national tragedies of the post-war era.'[39]

One of the most intemperate, though at times well-informed, indictments of the Eisenhower administration's handling of the satellite competition with the Soviet Union was published in September 1958 by the columnists Drew Pearson and Jack Anderson. They had little to offer on the Truman period apart from excuses. They too borrowed Lippmann's idea about the damage done by McCarthyism, but turned the point into a condemnation of the Eisenhower administration for failing to resist McCarthy effectively.[40] That could only be done by ignoring the facts that the hey-day of the witch-hunting senator from Wisconsin had been under Truman, from 1950 to 1952, and that he had finally met with his come-uppance under Eisenhower as a result of what one historian has called 'his fatal confrontation with the Army' in 1953 and 1954, at about the time that the American ICBM project was finally put into top gear.[41] Pearson and Anderson also adopted a partisan line of attack first devised by Senator Symington's staff, which tried to saddle Eisenhower with the responsibility for cancelling America's only ICBM project, then in its embryonic study stage, back in 1947, because he had been Chief of Staff at the time. The fact that the company involved, Convair, had to take its decisively unsuccessful appeal against the decision not to Eisenhower but to the Army Air Forces (AAF) Headquarters, as they themselves related on the following page, suggests that this was a somewhat flimsy charge.[42]

Another device used by Pearson and Anderson to allocate most of the blame for the sputniks crisis to the Eisenhower administration was the claim that Truman's intelligence apparatus never found any evidence of a major Soviet R&D effort in the missiles field, to which it would have been necessary to respond. Writing of Secretary of Defense Louis Johnson, they commented that:

> It should be noted in fairness that he can be blamed only for his failure to foresee Russia's rapid technological advance. The Eisenhower Administration had ample evidence of this advance.[43]

The inaccuracy of this comparison between the intelligence positions of the two administrations was almost certainly known to its authors, who had excellent sources in government, and particularly within the intelligence community. But such a comment can only be substantiated through a careful examination of the still incomplete information now available on American intelligence of Soviet rocket programmes before 1953, which will be offered in Chapter 5. At this point the breathtaking deceit of Pearson and Anderson can simply be indicated by the following remarks, made by an unnamed participant at the first symposium on space medicine in the United States, which was held on a semi-public basis at the US Air Force's School of Aviation Medicine, Randolph Air Force Base, Texas, in November 1948, several months *before* Louis Johnson became Secretary of Defense:

> It is now obvious that the Kremlin is going to develop the rocket as a long-range missile because they have moved to Russia most of the German rocket-producing personnel. Rockets have captured the imagination of the Russian militarist . . . Let us hope that their manifest attachment to rockets does not lead some of their predatory leaders to the idea that he who rules space will rule the world.[44]

If that was what Air Force scientists and their civilian colleagues were thinking and saying in November 1948, it is hardly credible that their country's intelligence service had not at least as good and probably a far better idea of the state of Soviet rocketry by that date.

In case it is felt that the disinformation or misapprehensions of such early space writers has little relevance to our present picture of the history of the space race, it is worth pursuing this example a little further. In one of the most influential histories of space exploration ever published (in 1966), Wernher von Braun and Frederick Ordway wrote of the early post-war period that:

> What little debate about artificial satellites that went on in the United States took place in a vacuum. Americans had the complacent idea that no one else was interested in placing a satellite into orbit. Little or no mention was made of the possibility that the Soviet Union was working on a satellite program. Russia was regarded as too backward technologically and too devastated by war to compete with the United States in any field.[45]

Since those lines were first written, relevant documents have been declassified which suggest that the picture of a Truman administration unfortunately deprived of any useful intelligence about Soviet rocketry is a highly misleading one. But the identical passage continued to appear in all subsequent editions of the book down to its most recent one in 1985.[46] Even so accomplished a space historian as Walter McDougall has recently ignored the available intelligence documents from the Truman period, preferring to endorse the traditional claim that Soviet progress in rocketry took American experts by surprise in about 1952–3, which of course implies that they knew nothing about it before then.[47]

However, it should be clear to the reader by now that such questions cannot be resolved without a painstaking examination of (still incomplete) historical evidence. Before proceeding, it seems advisable to offer a short overall account of the origins of American space policy, the details of which can then be taken up in subsequent chapters. Readers familiar with this story may prefer to pass immediately to Chapter 2, which concludes this introductory section by examining a topic which cannot easily be subsumed in later chapters, the awareness of rocketry and its implications for future astronautics which was gained by the two protagonists of the sputniks crisis, President Eisenhower and Senator Johnson, *before* 4 October 1957.

## 3   AN OUTLINE OF EVENTS

The events to be discussed in this book can be summarized as follows. By the end of World War 2 military planners in several countries had realized that a combination of the two most potent weapon technologies developed in that conflict, the atomic bomb and the long-range missile, would eventually become the key to military supremacy for a handful of the world's most powerful nations. In the Soviet Union an intensive effort to develop and eventually to marry both technologies was mounted in the late 1940s and early 1950s. In the United States, after a brief post-war lull, a vigorous programme for the further development and stockpiling of atomic and, later, thermonuclear weapons was also set in motion. But the effort to acquire long-range missiles, particularly of the more effective ballistic variety, was much less intense. This policy was maintained, despite the available intelligence about the early phase of the Soviet missile programme, virtually up to the end of the Truman administration, or in other words for well over half the twelve-year period between the end of World War 2 and the launching of the first sputniks.

The strategic situation of the two countries was very different. Soviet planners had strong reasons to acquire intermediate-range nuclear-armed ballistic missiles *en route* to developing the intercontinental variety.

Because of the greater concentration of American, as compared with Russian, centres of industry and population, they also experienced less disincentive from the combination of the relatively low explosive power of atomic bombs with the likely inaccuracy of first-generation intercontinental missiles. However, the widespread conviction of most senior US Air Force officers that for the foreseeable future their country's nuclear arsenal could be satisfactorily deployed and if necessary delivered by manned bombers also played a key role in delaying the start of a serious American ICBM project. Instead, long-range ballistic missiles remained the subject of paper studies only till after January 1953, when Eisenhower took office. Actual development was initiated by the Department of Defense early in 1954, some four or five years later than the Soviet equivalent.

During the Truman administration there were also numerous studies of the possible development and applications of artificial Earth satellites. At the RAND Corporation, a non-profit think-tank working for the US Air Force, this work reached a fairly advanced stage of design for a military reconnaissance satellite. But RAND's early recommendation, that a satellite project should be initiated promptly in order to ensure that the United States would be first into space, also went unheeded.

In the early 1950s the International Council of Scientific Unions (ICSU) accepted a proposal from a group of American scientists to sponsor a period of international cooperation in the geophysical sciences in 1957 and 1958, entitled the International Geophysical Year. The suggestion that the hitherto untried technology of artificial Earth satellites should be inaugurated in the context and for the purposes of the IGY was not publicly adopted, however, until October 1954, by which point a considerable gap had already opened between the levels of the necessary rocket technology in the Soviet Union and the United States, even though this fact was not widely understood in the latter. In the following months both the Soviet and the American Academy of Sciences set up special committees to examine the feasibility of a scientific satellite project. In 1955 the United States announced that it would attempt to launch satellites for the IGY, and a similar commitment was made a year later by the Soviet Union. An advisory group of American scientists boldly recommended that a new three-stage rocket, proposed by the US Naval Research Laboratory, should be designed, built, tested, and used as an IGY satellite-launcher within a mere eighteen months. But in fact this system, the Vanguard, did not succeed in placing one of its full-size satellites in orbit until shortly after the end of the IGY. Meanwhile the Russians launched their first three sputniks in October and November 1957 and May 1958, each time placing very large amounts of mass in orbit by comparison with early

American systems. In December 1957 the first American satellite launch attempt, a Vanguard carrying a small 'test sphere', failed catastrophically on the launch-pad in front of the cameras of the world press. After delays with a second Vanguard attempt an alternative US Army rocket system, Juno 1, succeeded in launching the first American satellite, Explorer 1, on 31 January 1958. It was followed in March by the first successful small Vanguard. By the end of 1958 the United States had launched one non-IGY and four IGY satellites, and the Soviet Union had launched three IGY satellites. But even allowing for the greater miniaturization of American payloads, the Soviet Union had placed far more scientific apparatus in space, including of course the pressurized cabin, life-support systems, and biological monitoring apparatus for the dog on Sputnik 2.

After two key specifications had been issued by the US Air Force in November 1954 and March 1955, the secret American project on reconnaissance satellites was finally upgraded from the study to the development phase in April 1956. After many setbacks, an experimental version of the system was first demonstrated in August 1960 and regular operational launches soon followed. Comparable Soviet military satellites began to be placed in orbit two years later, since when large numbers have been launched by both countries for a variety of military and civilian missions.[48]

# 2 *Ad Homines* – Eisenhower and Johnson Before 1953

Historians have often commented on the roles played by President Eisenhower and Senator Johnson as the two protagonists in the sputniks crisis of 1957–8, usually to the detriment of the former and the credit of the latter. But none has ever prepared the ground for such a comparison by asking whether each man's previous career made it *likely* that he would fail or pass the test of the sputniks by quite so decisive a margin as is usually claimed. The record of the Eisenhower administration on missiles and satellites, prior to Sputnik 1, will be examined below in its proper place, as will Johnson's handling of the political opportunity with which he was presented in the fall of 1957. This chapter deals only with what is known about the two men's previous encounters with the technology and its strategic implications.

## 1 EISENHOWER

Dwight D. Eisenhower was initiated into the field of what were then regarded as long-range missiles while Allied Supreme Commander at the end of World War 2. The German V-weapon bombardment came near the end of preparations for Operation Overlord, the D-Day landings on the coast of Normandy. As he later assessed it:

> It seemed likely that, if the German had succeeded in perfecting and using these new weapons six months earlier than he did, our invasion of Europe would have proved exceedingly difficult, perhaps impossible. I feel sure that if he had succeeded in using these weapons over a six-month period, and particularly if he had made the Portsmouth-Southampton area one of his principal targets, OVERLORD might have been written off.[1]

In April 1944, according to the official history, Eisenhower overruled the objections of senior officers of the US Army Air Forces in Washington and ordered that top priority be given to counter-attacks against suspected

German missile launching sites. But getting this actually to happen was to prove an uphill struggle. The British Air Ministry and Bomber Command were reluctant to comply, despite the urgings of Air Chief Marshall Roderic Hill, commanding the Air Defence of Great Britain. On 16 June 1944, just after the V-1 cruise missiles came into action (three months before the V-2 ballistic missiles), Eisenhower therefore repeated his order that such targets should:

> . . . take first priority over everything except the urgent requirements of the battle; this priority to obtain until we can be certain that we have definitely gotten the upper hand of this particular business.[2]

At the end of August 1944 the Allied Chiefs of Staff even went so far as to ignore months of excellent intelligence on the German A-4 (V-2) rocket project, notably assisted by the Swedish government's gift of the remains of a test missile which had strayed across the Baltic to land near Malmö on 13 June, by declaring 'that rocket attacks on London need no longer be expected'.[3] A few days later the first V-2s landed on Paris and London. Even after the new ballistic weapon had been in action for several weeks, with a single missile causing 264 casualties at New Cross in late November, Hill was unable to persuade his colleagues to release more resources for offensive counter-attacks. These facts do not, however, support the view of some historians that Eisenhower was personally responsible for the alleged decision 'to give only a low priority to the bombing of rocket-sites'.[4]

*        *        *

During the post-war scramble amongst the Allies for German secret weapons Eisenhower supported American technical intelligence teams attempting to transfer German rocket experts and hardware to the United States through Operation Overcast (later renamed Paperclip) and other missions. His cables to Washington requested firm policy decisions in favour of such acquisitions, describing them as 'the only reparations we are likely to get' and 'about the only material dividend we are likely to get from the War'.[5] When the British protested that a shipment of 341 rail-car loads of V-2 parts seized by the US Army Ordnance mission was in breach of the wartime agreement between the two countries, that all such material be shared equally, Eisenhower did intervene to stop the shipment, but acted too late to prevent it. His subsequent ban on further shipments was largely cosmetic, since the Army Ordnance mission had by then abandoned the Nordhausen factories to the Russians.

Eisenhower became Chief of Staff at the US War Department in

November 1945. In March 1947 he was briefed, rather poorly, on the methods and rationale of Operation Paperclip, which by then was meeting with both public criticism and determined administrative obstruction, most notably from within the State Department. Again, it received his full support.[6]

During Eisenhower's term as Chief of Staff the Army Department of Intelligence (G2) repeatedly assessed the value of the Paperclip rocket scientists alone 'at hundreds of millions of dollars' and as saving 'two to ten years in American research'.[7] This was also the view of the Army Air Forces (AAF). J. Edgar Hoover, director of the Federal Bureau of Investigation, considered the imported Germans to be a major security risk, and urged that they be refused all contact with the military research establishments to which they had been assigned. But in June 1946 Eisenhower made it plain that he shared the views of the intelligence departments:

I cannot overemphasize the necessity for the maintenance of a well-rounded and vigorous research and development program. . . . In the fields of guided missiles, electronic devices, and high-speed, high-performance aircraft the possibilities are unlimited. It would be fatal to permit ourselves to be outdistanced.[8]

He repeated the message even more firmly in February 1947:

In the field of guided missiles, electronics, and supersonic aircraft, we have no more than scratched the surface of possibilities which we must explore in order to keep abreast of the rest of the world. Neglect to do so could bring our country to ruin and defeat in an appallingly few hours. Those of us who were in Europe in the black days when Hitler was making his last desperate gamble with the V-1 and the V-2, know how close to success that gamble came. Yet those weapons, terrible and effective as they were, were child's toys in comparison with those which can be produced.[9]

However, Eisenhower never shared the 'utopian' attitude to such weapons of some of his subordinates in the AAF. The remarks just quoted were preceded by a reservation:

I am inclined to deplore, to some extent, the statements of the 'push-button warfare' or what is sometimes called the Buck Rogers school of thought, concerning our future requirements in this [R&D] field. Such thinking, if followed too far, works against our efforts toward building the balanced force which I am convinced will be required during the lifetime of all of us here.[10]

In February 1951, speaking now as NATO's first Supreme Allied Commander, Europe (SACEUR), Eisenhower testified again to the same effect:

> Gentlemen, wars have never been won by any one particular weapon or any particular gadget. . . . there is a combination of power that is applied against your enemy that is calculated to destroy his will to exist; his will to carry on the battle . . . [11]

Nevertheless, when the Missouri industrialist John Olin joined the 'veritable stream of American visitors passing through my office'[12] to bring information about the advances being made with guided-missile technology at the Swiss Oerlikon Company, Eisenhower thought the matter important enough to draw to the attention of the then Army Chief of Staff, General Lawton Collins.[13]

In general, Eisenhower seems to have maintained an intelligent interest in new military technologies. As we shall see below, there was a slight but growing awareness at this time, in some parts of the American armed forces, of what missile programmes might eventually mean for astronautics. It is therefore no surprise that, well before he publicly endorsed an American satellite project for the IGY as President, it was being said that 'Dwight Eisenhower is reported to have stated his belief (while still a general) in the future of space travel.'[14]

*          *          *

In view of the attention to be paid to such matters below, it is also appropriate here to notice Eisenhower's marked interest in the psychological and propaganda aspects of warfare and international conflict, including those of a covert or semi-covert nature. In 1943 General Dwight Eisenhower and his brother Milton welcomed the establishment of a Psychological Warfare Board (PWB) under C. D. Jackson, the editor of *Fortune*, and its integration into the work of the Allied Forces Headquarters in North Africa and Italy.[15]

After the war Eisenhower's interest in psychological warfare, his keenness to strike back at the Communist enemy, and his respect for the ideas of prominent cold warriors such as Henry Luce and C. D. Jackson, led him to associate with some of the activities of the National Committee for a Free Europe and the Crusade for Freedom. In 1949 the former body, under Jackson's leadership, set up Radio Free Europe to broadcast Western propaganda to the 'satellites' of Eastern Europe. In 1951 it established Radio Liberty to do the same for the Soviet Union. The chairman of Crusade for Freedom was Lucius Clay, formerly one of Eisenhower's generals and military governor of the American occupation zone. Eisenhower was a member of the board, and helped with fund-raising activities. However he disapproved of the Crusade's plan to launch thousands of propaganda balloons across the Iron Curtain, which it eventually began doing on 15 August 1951. Since he was already being sounded as a potential candidate for the presidency by various constituencies it was only prudent for him to avoid associating himself with activities that could be thought illegal or needlessly provocative.[16]

In January 1951 a State Department paper on the psychological exploitation of Eisenhower's appointment as SACEUR was forwarded to his Chief of Staff at SHAPE, Lieutenant General Alfred Gruenther.[17] It seems to have struck a chord, for a few days later Eisenhower is said to have shouted down a chorus of complaints about shortages at his first NATO planning conference by declaring that 'I know there are shortages, but I myself make up for part of that shortage – what I can do and what I can put into this.'[18] Later in the year he wrote to Robert Lovett, the newly appointed Secretary of Defense, that what Europe needed, and he intended to supply, was 'a Moses'.[19]

In February 1951 Eisenhower summed up his emphasis on psychological warfare in testimony to the Senate:

> As I tried to make clear this morning, we are trying to generate a morale. I conceive that to be the real duty I owe the United States. . . .
>
> The minds of men are what we are trying to capture, and to get better and stronger allies, reliable allies. We can do it, for example, in these countries whose help we want so firmly. We can do very much by getting the truth in there, *not as a government thing by us. It has to be by people that are clever enough to do it clandestinely.* [The italicised passage was suppressed in the published version.]
>
> This applies behind the Iron Curtain, in front of the Iron Curtain, everywhere in the world.[20]

## 2   LYNDON B. JOHNSON

Lyndon B. Johnson was elected to the House of Representatives in 1937 and at once began to concern himself with military affairs, partly as a result of an unusual recommendation from President Roosevelt which led to his immediate appointment to the Naval Affairs Committee.[21] During World War 2 Johnson headed a special investigating subcommittee of that body which dealt with such matters as procurement policy, contracts, and absenteeism at armaments factories. After fighting an election campaign on the themes of 'Peace, Preparedness and Progress' he entered the Senate in 1949 and joined the Armed Services Committee under the chairmanship of Richard Russell. During that year he allied himself closely with Stuart Symington, then Secretary of the Air Force, in attempting to resist the cuts in aircraft procurement that formed part of an overall economy drive initiated by the new Secretary of Defense Louis Johnson. However, most senators upheld the cuts, and effective opposition came largely from the House of Representatives.[22]

In February 1950 Senator Johnson addressed his colleagues on the issue of guided missiles:

> Our missile program is a minor, even obscure, item in the defense budget. Each of the three services has a guided-missile program of its own – but, as yet, no really worthwhile guided missiles. Publicity about some of our rocket research has created a largely false impression that we have missiles which would be used in the defense of this country. As for now, we have none.[23]

Accepting that a climate of economic stringency obtained, however, he confined himself to the politically safe, because impractical, suggestion that funds should be transferred to the missile programme at the expense of other items.

In July, shortly after the outbreak of the Korean War, Johnson was chosen by Russell and other senior senators to chair a new Preparedness Investigating Subcommittee of the Armed Services Committee. Its first investigation revealed major shortcomings in the Munitions Board's arrangements for the production of synthetic rubber. Others dealt with price fixing in the tin market, inefficient uses of manpower in the Army and Air Force, substandard housing for military families, and the exploitation of servicemen through illegal slot machines. One inquiry which Johnson chose to direct in person looked into the procurement of military ordnance, a field which included tanks, bazookas and a 'long-range projectile' programme.[24] Like the rest of the Subcommittee's work, this seems to have been addressed mainly to the immediate needs of the American forces in Korea. Its practical results are obscure, but sixteen months later one of Johnson's aides did point out that the average delay on guided missile deliveries to the armed forces was still 46 per cent.[25] Johnson himself subsequently claimed that at this time 'I annoyed several officials in the Pentagon by pointing out that they were paying too little attention to guided missiles.'[26]

In December 1950 Johnson wrote in a Houston magazine that:

> . . . we will have to spend more time and effort, I believe, on developing radically new weapons of war – principally guided missiles. The old-style warfare in Korea should not lower our sights from the goal of inventing new weapons that might revolutionize wars – and keep us ahead of our enemies. We should pool our best scientific brains on guided missile research, just as we pooled our best brains to develop the atomic bomb. Guided missiles offer our best hope of a successful answer to sneak air attacks. Such instruments might become powerful deterrents against aggressors.[27]

During the remainder of the Truman administration the Preparedness Subcommittee continued its investigations of such needful aspects of military strength as the supplies of tungsten, tires, meat, nickel, tin, steel and rubber, examined the production of aircraft, and highlighted such instances of waste as the use of 200 lb of coffee for sweeping floors at Carswell Air Force Base.[28] In November 1951 Senator Johnson urged his Democratic colleagues to adopt Eisenhower as their presidential candidate. In October 1952 he said that new weapons would soon begin to alter the character of war. In January 1953 he took on the demanding role of Democratic Leader in the Senate, and due to the Republican majority in the Senate for that session he lost his chairmanship of the Preparedness Subcommittee during 1953 and 1954 to Senator Estes Kefauver. He regained it in 1955, but became at the same time the youngest Senate Majority Leader in history, with a margin of only a single seat in favour of his party.

Perhaps as a result of these developments Johnson took little interest in Eisenhower's guided missiles programme before 4 October 1957. Thus a list of twenty-one major projects in hand at the Preparedness Subcommittee in January 1957 includes a variety of items from a sleeping-bag contract to longitudinal instability in the F-101 fighter aircraft, but none relating to missiles. And when a member of Johnson's staff suggested at the beginning of February that the Subcommittee should mount an enquiry into the missiles programme, to be headed by Senator Symington, his idea was not picked up.[29]

*     *     *

It is instructive to make a careful comparison between Eisenhower's and Johnson's published remarks about missiles during the Truman administration. Johnson emphasized the potential of missiles 'in the defense of this country' and as 'a successful answer to sneak air attacks'. As will be shown below, this was the conventional wisdom of the Truman administration, which thought of missiles primarily as a short-range armament for strategic defence against bomber attacks. Eisenhower, for all his reservations about any single type of weapon ever becoming decisive by itself, realized early on that missiles had awesome technical potential as *offensive* weapons, and might indeed become so much more 'terrible and effective' than Hitler's V-weapons that they could threaten the 'ruin and defeat' of an entire country (sc. many hundreds or even thousands of miles away from their launch sites) within a few hours.

When it comes to applications of rockets for space technology, the comparison between the two men's early thinking stumbles for lack of relevant material. Neither first-hand witnesses, nor the man himself, nor historians have ever claimed that Johnson was interested in space policy before the sputniks. Eisenhower too was never a space enthusiast. But certain aspects of his career before the sputniks crisis pose a problem for historians which has seldom been appreciated. It seems most improbable, *a priori*, that he should have neglected space policy before the sputniks to the degree that is usually claimed. Those who take such a position should try to explain why such an uncharacteristic course of action was taken by a President who had long been aware of the strategic significance of rocket technologies, of the general importance of prestige for American foreign policy, and of the part played by scientific and technical achievements in upholding that prestige. They have never convincingly done so.

# Part II  Truman

# 3 Advice on Missiles for Truman

## 1 AMERICAN SCIENCE ADVICE IN THE 1940s

In order to form a clear picture of the part played by the Truman administration, by omission or commission, in the shaping of the early space policy of the United States, we must assemble a variety of historical evidence. We need to know what kind of advice was available to the administration about the technical feasibility and the military and other applications of both long-range missiles and artificial satellites. It is also helpful to review the development of any wider public awareness of the potential of such technologies. Next, we need to form the best possible estimate of the secret intelligence on Soviet rocketry that was provided to the Truman administration. These matters will be discussed in this and the two following chapters. Only after establishing the facts on the input side of policy can the output side, the policy itself on potential space technologies, be properly assessed. But before beginning with a review of the climate of public awareness and expert advice on missiles under Truman, it is necessary to say something about the early post-war development of science advice in the United States in general.

Science policy may be defined as the set of government decisions about what to do for, with and about science as a whole, with primary emphasis being placed on areas regarded as significant for national interests. But the domain of science advice comprises the endeavours of leading scientists to influence not merely science policy but government policies in general, with primary emphasis being placed on areas regarded as especially in need of scientifically expert clarification for policy-makers, or as especially urgent in the light of perceived professional, national, or, sometimes, humanitarian goals. The origins and development of the post-war structures for the provision of science advice to the United States government have been described by several of the protagonists as well as by historians.[1] All that will be attempted here is a review of factors on both sides of the relationship which may have influenced either the quality of the advice being offered in respect of potential space technologies or the degree of attention paid to that advice by the Truman administration, which after all governed the United States for well over half the period between the end of World War 2 and the launching of Sputnik 1.

As the first post-war administration Truman and his colleagues had the opportunity to lay down the basic principles of modern American science policy. By rejecting Congressional bills to establish first the Atomic Energy Commission (AEC) and later the National Science Foundation (NSF) as élitist and autonomous bodies, relatively insulated from direct control by the President, and by ensuring the passage of alternative measures constituting both institutions as executive agencies responsible to the President and his administration, like other parts of government, President Truman accepted and fulfilled that responsibility as he thought best.[2]

However, the legislative struggles surrounding the AEC and NSF delayed the emergence of anything like a comprehensive science policy. On the military side, the wartime Office of Scientific Research and Development (OSRD) was replaced by the Joint (Army and Navy) Research and Development Board (JRDB), renamed the Research and Development Board (RDB) when the new, 'unified' Department of Defense was created in 1947. These structures were headed by Vannevar Bush, a former electrical engineer from MIT who dominated military science advice in the United States from 1940 until his retirement in October 1948. But the RDB reflected the problems of the Department which it served, often operating as yet another interservice battleground. Other powerful science advice bodies, some of them formed in the immediate post-war years, were maintained as organs responsible only to the separate armed services. They included such important institutions, for the nascent American missile programme, as the Army Ordnance Advisory Committee, the Office of Naval Research (ONR), and for the Air Force, the Scientific Advisory Board (SAB) and the RAND Corporation. Separate structures also existed for the development of science advice on civil and military nuclear energy. Partly because of the perceived importance of their subject matter, and partly because they seemed to operate outside the squabbling services as truly 'national' institutions, the nuclear advisory bodies tended to dominate the channels between civilian scientists and government. Their cultural weight in the strategic debate, together with the inability of the administration to impose a comprehensive schema of scientific and technological priorities on the competing agencies, may well have contributed to the neglect of long-range missiles during this period.

In the immediate post-war years the administration combined a complacent neglect of the American nuclear arsenal with an emphasis on attempts to establish an international system for the control of all applications of atomic energy. In July 1949, however, the President decided that international control was not attainable, and that therefore 'we must be strongest in atomic weapons'.[3] At the same time physicists on the Atomic Energy Commission's General Advisory Committee (GAC) were going through the stressful process of deciding whether or not to recommend that the United States proceed to develop the hydrogen bomb. In early September American intelligence detected an atomic explosion which had occurred in the Soviet Union on 29 August; after a short delay the news was announced by the President. This development undermined the authority of previous assurances given by the GAC's chairman Robert Oppenheimer (the former director of the Los Alamos Laboratory at which the wartime atomic bombs had been built) that, when it came to 'obtain[ing] a significant atomic armament . . . for a long time to come the Soviet Union will not have achieved this objective'.[4]

The atomic physicists advising the administration were seldom neutral on strategic issues. Indeed, they were amongst the founders of the two opposing schools of thought which have contended for over four decades for control of the line to be followed in foreign and military policy by the United States. The first tends to seek cooperative, bilateral solutions to security problems, through negotiation with the Soviet Union. On the armaments side, they have usually favoured the deployment of relatively limited strategic force to 'contain' the Soviet Union by means of nuclear deterrence. The second prefers to stick to unilateral solutions based on decisive strategic superiority, which is usually sought from weapons technology. Their aspiration has therefore been to acquire the capability of inflicting massive nuclear devastation on the Soviet Union, while incurring a minimum of reciprocal damage to the United States. (A Strategic Air Command briefing in the early 1950s referred to a capacity to transform the Soviet Union into 'a smoking, radiating ruin at the end of two hours'.)[5] The advisers' responses to the first Soviet atomic test marked a major setback for the former group, which then included Oppenheimer and the AEC's president David Lilienthal, and the rise to political prominence of a group of conservative-minded physicists who were concerned at the neglect of the United States' atomic arsenal and convinced that a hydrogen or 'Super' bomb both could and should be developed as a matter of urgency. They provided AEC commissioner Lewis Strauss with the technical advice which enabled him and Senator Brien McMahon, chairman of the Joint Congressional Committee on Atomic Energy, to lead a successful campaign in favour of an H-bomb programme, despite the GAC's decision in October 1949 to recommend against it.[6]

As conservative scientists gained the upper hand in policy-making circles, President Truman took a series of decisions in the early part of 1950 to the effect that the H-bomb should be. vigorously pursued. But physicists of the more 'dovish' school of thought refused to give up. They continued putting forward ideas which they had been canvassing for some time about more limited and feasible uses of nuclear weapons, and about the need to establish the scientific and technical basis for such applications. One of their central themes in the late 1940s and early 1950s was that the United States should devote more effort towards acquiring capabilities for early warning of, and for active defence against, strategic nuclear attack. Two prominent advocates of this point of view were Lloyd Berkner and Robert Oppenheimer.

After the outbreak of the Korean War in June 1950 the Department of Defense greatly expanded the practice, with which it had already experimented, of contracting with *ad hoc* groups of civilian scientists for studies of military issues in which technical factors played a dominant role. The advice generated by the rash of 'summer studies' indicated the feasibility and desirability of developing more flexible offensive and more capable defensive systems, a 'limited containment' approach which ran counter to the established preference of the administration and the dominant service, the US Air Force, which was for a strategy of deterrence by means of an overwhelming offensive superiority based on the manned bomber. One such study was Project Vista, a response to recent technical developments which had shown that comparatively low-yield atomic weapons were now possible. Vista was conducted at the California Institute of Technology (CalTech) under the direction of its president Lee DuBridge, another GAC member, from April 1951, and presented its report a year later. It was primarily an investigation of the military implications of small or 'tactical' atomic weapons, and it coincided aptly with Senator McMahon's initiation, in September, of political moves to secure a major expansion in the production of atomic weapons and in the range of types produced.[7]

In the 1940s Lloyd Berkner had alternated between various forms of government service, including a year as executive secretary of the JRDB with Vannevar Bush, and scientific work at the Carnegie Institution's Department of Terrestrial Magnetism (DTM). But by 1951 he was president of Associated Universities Incorporated, an organization established by various East Coast universities in order to create and manage the Brookhaven National Laboratory, which contracted with the government for a variety of nuclear research. In that capacity he directed the East River Study on civil defence at the Massachusetts Institute of Technology (MIT). Completed early in 1952, the East River report concluded that civil defence measures would be largely ineffective without major improvements in early-warning systems and active anti-aircraft defences, and urged that these should now be intensively developed.[8] Project Charles, a parallel study of air defence with some Canadian and British participation, also began at MIT in January 1951. The director was Frank Loomis, and amongst those taking part were Lloyd Berkner, Louis Ridenour, Charles Lauritsen, Edwin Land, Guyford Stever, John von Neumann, and Jerrold Zacharias.[9]

Project Charles helped to motivate the establishment of MIT's Lincoln Laboratory in September 1951, with Loomis as its first director. Funded largely by the Air Force, the Laboratory's field of research was to be the scientific and technical problems of air defence, and particularly of radar. A further special effort on the problems of air defence was the Lincoln Summer Study of 1952, which brought physicists from all over the country for intensive discussions at the new Laboratory over about four weeks. It was directed by Zacharias, who later told the Oppenheimer security hearings that Berkner and Lincoln's Malcolm Hubbard, both veterans of Project Charles like himself, were among 'those of us who were involved full time besides myself'.[10] Albert Hill, who served as assistant and then deputy director at Lincoln, before taking over as director in July 1952, and was the senior member of the Laboratory with responsibility for the Summer Study, testified that Berkner's regular visits to MIT in early 1952, 'talking about the possibility of early warning', had been one of the factors leading to a growing scientific interest in this subject.[11] And Oppenheimer himself pointed out that he had supported the Summer Study and had taken a small part in it, as had Lauritsen.[12]

The scientists' panels of 1951 and 1952 were an important stage in the process whereby politicians, government officials, scientists, and military officers established a more enduring framework for the collaborative determination of strategic issues than had been erected during the early part of the Truman administration. This was so in spite of Air Force resistance to the 'finite containment' thinking favoured by many participants. They were also significant for the degree to which they emphasized areas of scientific investigation that were to be particularly relevant to early space research. (These will be described in Chapter 4.) By the summer of 1952, however, it was too late for the outgoing Truman administration to adopt any of the scientists' recommendations, a situation which Oppenheimer in particular found frustrating.

\*        \*        \*

The Truman administration was also advised of the importance of gathering scientific and technical information from other countries. Shortly after World War 2 the US Navy established a London branch of the ONR which had this as one of its main purposes.[13] Some years later, in 1952, the Air Force followed suit by setting up the Advisory Group for Aeronautical Research and Development (AGARD) at NATO's Paris headquarters, though not without meeting stiff bureaucratic resistance from the Navy.[14]

In 1949 an International Science Policy Survey Group was formed in the Department of State. The Group was supervised by an International Science Steering Committee chaired by Lloyd Berkner. A blue-chip advisory committee comprising such luminaries as Bush, Isidor Rabi and the chairman of the National Research Council, Detlev Bronk, also contributed to its work. In May 1950 a report on *Science and Foreign Relations* was published over Berkner's signature.[15] It called for increased interaction and cooperation with the scientists of other countries, in order to further both the political aims of the United States and also the scientific basis on which American security was now perceived as resting. As Gregory Comstock of the Stevens Institute of Technology later suggested to Truman's press secretary, Joseph Short:

> The Berkner Plan of the State Department . . . could be made a collecting agency for European know-how (nothing to pass over lightly) for distribution by the Office of Technical Services.[16]

However, the report's rather grandiose proposals for a network of science offices within United States embassies around the world went largely unfulfilled, and even the reorganized science policy apparatus within the State Department fell into disuse.[17] This may have owed something to the fact that shortly after submitting it Berkner began to switch the focus of his internationalism towards the International Geophysical Year, as did his fellow geophysicist J. Wallace Joyce, who had directed the Science Policy Group.

Several of these aspects of science advice during the Truman administration will be referred to again below. At present it is time to look more closely at the development of public awareness and expert advice about long-range missiles, whose rocket propulsion systems were one of the key technologies for early space missions.

## 2  PUBLIC AWARENESS OF MISSILES

One of the earliest notifications of the military potential of the long-range, rocket-powered ballistic missile was placed before the American public in 1931 by David Lasser, founding president of the American Interplanetary (later 'Rocket') Society (ARS). Lasser warned that defence against such missiles would be impossible, but that:

> Short of the cessation of warfare, there is no apparent way of barring [their] use. The rocket of 1950 may then be an agent of great good and of equal harm.[18]

In the 1940s speculation was supplanted in the public eye by the actual wartime experience of German rocket-powered weapons, of which the most effective were the unmanned flying or 'cruise' V-1 missile and the ballistic V-2 missile.

The title of 'Father of Space History' may be justly conferred upon Willy Ley, a founder member of the German Verein für Raumschiffahrt (Society for Space Travel) who emigrated to the United States in 1935. His first book in English on the subject was *Rockets*, published in May 1944. The first six chapters, well over half the text, were devoted to a history of rocketry up to the closing down of the German Society by the Nazi authorities in 1933. The book was written without access to Allied intelligence, but wartime applications of rocketry were mentioned briefly in an appendix. Unfortunately Ley rejected German claims about 'a gigantic long-distance rocket supposed to carry more than 100 miles' as propaganda, and argued that rocket missiles only made sense 'up to about 5000 yards'.[19] The first operational V-2 was fired on 6 September 1944; a second impression of Ley's book appeared in the same month.

By the end of the year, months before VE Day and longer still before the first explosions of atomic bombs were followed by the Japanese surrender, Britain's Assistant Director of Intelligence (Science), Reginald Jones, wrote in the magazine of the US Eighth Air Force that:

> With a very long-range rocket we may have to accept errors, and it may be easier to increase the radius of destruction by the use of new types of explosive based on the fission of the uranium nucleus. If such an explosive becomes practicable, it will probably have a radius of destruction of the order of miles, and on this account alone it might best be carried in some unmanned projectile . . . [20]

In January 1945 a revised edition of Ley's book brought detailed information on the world's first long-range ballistic missile to American civilians.[21] Further popular accounts of the German technical achievements and the future potential of long-range military rocketry soon followed.[22] In November Bernard Brodie, a civilian strategist who had worked for the US Navy during World War 2, declared that the combination of the atomic bomb with rockets 'capable of several thousand miles of range' was likely to become a militarily and economically viable weapon. He added a caveat that 'the problem of controlling the flight of rockets over such distances is greater than is generally assumed'.[23]

As early as June 1945 the US Army publicized the results and future potential of German weapons development programmes at a widely reported press conference in Paris.[24] Apprehension about the possible threat to the United States from long-range bomber aircraft armed with atomic bombs was quick to form, together with concern that defences against such an eventuality should be developed and deployed as soon as possible, if a reliable system for the international control of atomic energy could not be constructed. As Brodie pointed out, even if the best performance of anti-aircraft defences in the recent war – typically, about 10 per cent attrition against massed bomber attacks – could be doubled or trebled, it would have little strategic significance against large-scale atomic bombardment.

In his 1945 paper Brodie noted but dismissed a claim emerging from the House Naval Affairs Committee that 'an effective counter-measure to atomic bombs' had been devised, in the form of a remote detonator.[25] In a revised version he also referred to President Truman's attempt to allay anxieties about a future atomic threat to the United States by claiming, in an address to Congress on 23 October 1945, that 'Every new weapon will eventually bring some counterdefense against it.' Brodie responded that, whatever might be achieved in anti-aircraft defence, the penetrative qualities of the V-2 rocket meant that offensive armaments would continue for the foreseeable future to be overwhelmingly superior to defensive ones, with revolutionary consequences for warfare and strategy. He added that in World War 2 'the defenses against the V-2 rocket were of practically zero effectiveness, and those who must know about it admit that thus far there has been no noteworthy progress in defenses against the V-2'.[26]

Brodie's 1946 book was not the first open publication in which this problem was aired. In the 1945 Christmas supplement of *The Nation*, for example, Ivan Getting, an electronics engineer from the Radiation Laboratory at MIT with wartime experience of British attempts at anti-missile defence, had laid the matter before at least a section of the general public.[27] And early in the New Year an anthology on atomic weapons included an article by the physicist Louis Ridenour arguing that 'There Is No Defense' against a nuclear missile attack, while omitting to point out that there was, as yet, no threat of one either.[28]

At about the same time Secretary for War Robert Patterson misleadingly announced that the German scientists brought to the United States in

Operation Paperclip had already helped to develop more effective missiles than the V-1 and V-2.[29] His claims appeared to be supported by some of the Germans interviewed in *Time, Life*, and *Newsweek*, in a public-relations move on behalf of Paperclip which largely backfired.[30] It was perhaps an American cultural inclination to take the technological wish for the hard-won engineering deed which encouraged a widespread belief that long-range rockets, which fourteen years earlier had been merely a gleam in the eyes of a handful of enthusiasts, were now just around the corner. The military implications for the nation which had initiated 'the atomic age' were sobering indeed.

By September 1947, when the columnists Joseph and Stewart Alsop asked whether the United States was prepared for a 'push-button' war, few readers would not have realized from the title alone that the subject to be discussed was long-range missile technology.[31] The Alsops publicized an estimate given to Congress earlier that year by Lieutenant General Ira Eaker, Deputy Commander of the Army Air Forces, that intercontinental ballistic missiles (ICBMs) were between ten and eighteen years off.[32] In October *The Atlantic Monthly* explained that Eaker's estimate had been for a programme conducted at a peacetime pace, and that he had also said that a crash programme could produce ICBMs within five years.[33] The significance of these alternatives had already been noted, with characteristic prescience, by Brodie:

> ... in so far as the development of rockets nullifies that type of disparity [air superiority] in offensive power, it should be noted that the development of rockets is not likely to proceed at an equal pace among all the larger powers. One or several will far outstrip the others, depending not alone on the degree of scientific and engineering talent available to each country but also *on the effort which its government causes to be channeled into such an enterprise.* In any case, the possibilities of an enormous lead on the part of one power in effective use of the atomic bomb *are inseparable from technological development in vehicles* ... [34]

## 3 EXPERT ADVICE ON MISSILES

Before World War 2 research on rocketry was relatively neglected in the United States by comparison with the situation in Germany or the Soviet Union. The American rocket pioneer Robert Goddard worked in isolation at first, and chose to continue doing so even when approached by members of the ARS with a view to cooperation.[35] At the main governmental agency for aviation research, the National Advisory Committee for Aeronautics (NACA), 'it would have been quite impossible in the prewar period to have any major support from the military, industry, or from Congress for research and development aimed at such radical concepts as . . . the rocket engine . . . '[36] At the Guggenheim Aeronautical Laboratory of the California Institute of Technology (GALCIT) a handful of graduate students of the great aerodynamicist Theodore von Kármán began unofficial rocketry experiments in the late 1930s. Their work resulted in a limited development programme during World War 2, which was at first confined to Jet Assisted Take-Off (JATO) applications. In the last twelve months of the war the GALCIT rocket project, renamed the Jet Propulsion Laboratory (JPL), began research on missile applications which eventually led to the deployment of the United States' first operational ground-launched atomic missile, the Corporal, in 1951, as well as to early successes with a vertical 'sounding' rocket, the WAC-Corporal.[37]

The end of World War 2 was a turning-point in the history of American rocketry. In May 1945 the commander of the German missile development station at Peenemünde on the Baltic coast, Major General Walter Dornberger, together with the station's technical director, Wernher von Braun, and several hundred other technical experts, surrendered to American forces in the Bavarian Alps, where they had taken refuge from the advance of Soviet forces in the north of the country. After several weeks of interrogation over 200 of these key personnel were shipped to the United States under Army contracts and without immigration formalities, through Operation Paperclip. Shortly before the final constitution of the Soviet zone, US Army Ordnance Technical Intelligence teams also took possession of the bulk of V-2 rockets and rocket parts at the Nordhausen factory, and of all the technical documents concealed in an abandoned iron mine nearby.[38] This material, and even more so the German experts who had created it, were eventually to play a leading role in the American space programme.

*                    *                    *

In November 1945 General 'Hap' Arnold, commander of the US Army Air Forces, published an open *Third Report to the Secretary of War*. In it he concluded that long-range missiles could probably replace bombers in the foreseeable future:

> Improvements in aerodynamics, propulsion, and electronic control will enable unmanned devices to transport means of destruction to targets at distances up to many thousands of miles. . . .

When improved air defenses make [manned long-range atomic bombing] impracticable, we should be ready with a weapon of the general type of the German V-2 rocket, having greatly improved range and precision, and launched from great distances. V-2 is ideally suited to deliver atomic explosives, because effective defense against it would prove extremely difficult.[39]

Arnold's thinking about 'a manless Air Force' grew out of a long-standing enthusiasm for advanced aviation technologies; it was also supported in some respects by the findings of the Scientific Advisory Group (SAG; later Scientific Advisory Board – SAB) which he set up under Theodore von Kármán with instructions to tell him what 'the next Air Force' should be like.[40] The SAG surveyed the wartime achievements in aeronautics and rocketry of all combatants, but particularly those of Germany, and reported its initial findings in August, its recommendations in December 1945.[41]

In the 1940s there were, however, missiles and missiles. Specifically, there were guided and unguided or 'free' missiles. For the most part, the former were thought of as air-breathing and aerodynamic cruise missiles, like the German V-1, the latter as ballistic missiles, like the V-2, which would be thrown up to the edge of space by a powerful rocket which ceased to fire early in the trajectory, thereby removing the most obvious, and in the V-2 the only, basis for control systems.

The long-range guidance problem consists of three elements: what to 'tell' the missile, how to do so, and how to design a missile which responds appropriately. Discussion of the first two aspects will be deferred to Chapter 4. On missile design, the consensus of American experts in the 1940s was that purely ballistic missiles could not be made accurate enough against targets of appropriate size at intercontinental ranges. It was all very well for the Alsops to suppose from behind a typewriter that missiles soaring to a height of 'two hundred miles', 'at speeds up to those of meteors', could 'somehow be guided precisely, continuously and with exact control', provided that one had first been able to 'pass from the simple requirement of mechanical miracles to the need for actual triumphs over Nature'.[42] They had nothing to say about where such triumphs were to come from in the case of an unpowered projectile passing through a virtual vacuum above the (then underestimated) limits of the atmosphere. (But many of the scientific problems which had to be faced in the development of long-range ballistic missiles were linked with others related to the technologies of early warning and air defence which began to engage the attention of physicists in the summer studies of the early 1950s.)

Those who doubted the near-term military feasibility of long-range ballistic missiles occupied key positions. In his memoirs von Kármán claimed
that his initial report in August 1945 had included a 'key finding that
present technical competence could build ballistic missiles of 6000-mile
range, enough to blitz any country of the world from bases in any other
country'.[43] However, the reference to an indiscriminate bombardment, or
'blitz', was a veiled concession to the severity of the guidance problem.
When the Group's full 33-volume report was presented in December, after
Arnold's, the ICBM was no longer its 'key finding' – if it ever had been.[44]
Instead:

> The report spoke of V-2–type rockets, saying that it might be possible
> to increase their range by 30 times, and it spoke of a satellite as a
> 'definite possibility'. But there was no development of either thought.
> Once mentioned, both thoughts were dropped in favor of a stress on
> what could be done within the atmosphere.[45]

Contemporary testimony by the director of OSRD, Vannevar Bush, kept
closely to the line of the SAG report:

> *Bush*: My point is simply that we have plenty enough to think about that
> is very definite and very realistic – enough so that we don't need to step
> out into some of these borderlines which seem to be, to me, more or less
> fantastic.
>      Let me say this: There has been a great deal said about a 3000-mile
> high-angle rocket. In my opinion such a thing is impossible and will be
> impossible for many years.
> *Senator Tydings*: But not ultimately impossible?
> *Bush*: I say nothing is ultimately impossible because I do not know. But
> I can say with confidence that I would not know how to do it today, and
> I do not think that any man in the world, today, would. I am speaking
> about a 3000-mile rocket. . . .
>      But the people who have been writing these things that annoy me
> haven't been talking about that [a 2000-mile cruise missile]. They have
> been talking about a 3000-mile high-angle [sc. ballistic] rocket, shot
> from one continent to another carrying an atomic bomb, and so directed
> as to be a precise weapon which would land on a certain target such as
> this city.
>      I say, technically, I don't think anybody in the world knows how to
> do such a thing, and I feel confident it will not be done for a very long
> period of time to come. . . .

I have just been criticizing the report of General Arnold of the Army Air Forces. . . . [46]

When Senator McMahon referred to a recent article about the German wartime A-10 project for an intercontinental missile by none other than the Commanding General of the Army Air Forces, General Carl Spaatz, Bush continued:

> If you were talking about 400 miles or 500 miles, I would say by all means. That is what the Germans did with their V-2. I would say yes, even with 2500 miles.
>
> But 3000 miles? That is not just a little step beyond. It is a vastly different thing, gentlemen. I think we can leave that out of our thinking. I wish the American public would leave that out of their thinking.[46]

Bush's scepticism about long-range atomic missiles was actually contradicted elsewhere in the same hearings by such distinguished scientists as Harold Urey, Philip Morrison and Leo Szilard. Rear Admiral William Purnell, the Assistant Chief of Naval Operations (Materiel), also told the senators that the ideal method for the delivery of atomic weapons would undoubtedly be the rocket.[47] However, as will be shown in Chapter 6, the step-by-step approach to missiles recommended by von Kármán and Bush, in which ICBMs were regarded as an 'ultimate' and therefore low-priority goal, remained central to Air Force guided missile policy throughout the Truman administration.

\*                \*                \*

An additional factor often said to have counted against long-range ballistic missile projects in the 1940s was 'the preference of "blue-sky" air officers for manned bombers'.[48] The existence of such a prejudice is naturally hard to document, and few historians have tried to do so. One recollection of it was published in 1960:

> In the spring of 1947, Frank Davis, assistant to chief engineer Jack Irvine at the Downey plant, was sent to Washington and elsewhere to do what he could to stem the flow of rumors that Project MX-774 should be forgotten because of its 'extremely advanced nature'. . . .
>
> Davis found considerable opposition in [*sic*] intercontinental missile concepts. Within the Air Force itself, some diehard fliers were wondering what the air branch was doing in the missile game in the first place.

They contended that air vehicles without cockpits didn't belong in the Air Force.[49]

This story should be seen in context. When Davis went to Washington, 11 out of 28 Air Force missile projects had already been cut, in response to a sudden 25 per cent retrenchment imposed in December 1946, but interestingly not the Convair ICBM project (MX-774). Further cost studies showed that the reduced budget was insufficient for the remaining 17 projects. By May 1947 it had been reluctantly decided within the Air Materiel Command (AMC) that five (or six) more projects would have to go, among them the MX-774.[50] Thus, prejudice against long-range ballistic missiles may have influenced this decision, but it was enabled to do so largely by the 'dog eat dog' climate brought about by acute financial pressure.

*             *             *

Despite speculation about the possibility of using nuclear-powered rocket engines, American scientists in the 1940s simply failed to foresee the 400-fold improvement in mass ratios that was to be achieved between 1946 and 1959 by advances in chemical fuels, combustion technology, instrumentation, and air-frame construction.[51] For example, the Convair rocket had been designed around a hypothetical fuel with almost twice the best specific impulse then obtainable. In May 1947, when the head of AMC forwarded his recommendation that Project MX-774 be discontinued, one of his reasons was that 'the specific impulse of fuels which are available or promise to become available is too low'.[52] The consensus amongst experts and senior officials remained that atomic-armed ICBMs would not be seen before the 1970s.[53]

The debate between champions and critics of the idea of 'push-button warfare' was sometimes warm. After retiring from the post of director of the RDB in October 1948, Vannevar Bush wrote a book which implicitly rebuked officers such as Generals Eaker and Arnold for expressing their view that atomic ICBMs could and should be developed faster than the official timetable. Bush had accepted by then that 'there is no defense against the high-flying [i.e. ballistic] missile except to prevent its being launched', and continued:

We are hence decidedly interested in the question of whether there are soon to be high-trajectory guided missiles of this sort spanning thousands of miles and precisely hitting chosen targets. The question is particularly pertinent because some eminent military men, exhilarated perhaps by a short immersion in matters scientific, have publicly asserted that there are. We have been regaled by scary articles, complete with maps and diagrams, implying that soon we are thus all to be exterminated, or that we are to employ these devilish devices to exterminate someone else.[54]

Bush proposed to 'treat' decisively such 'prognostications of doom . . . pulled from the Pandora's box of science, often by those whose scientific qualifications are a bit limited'. But he now agreed that, with the aid of atomic explosives to compensate for its relative inaccuracy, the intercontinental missile could 'conceivably enter in time', although its guidance systems would never perform more than 'nearly as well as a manned aircraft'. His argument against the ICBM now turned mainly on costs. Since, on his assumption, chemical rocket fuels could not be further improved, missiles would be very large and therefore expensive. Atomic warheads would continue to be a scarce and costly resource – a second rash supposition which ignored Brodie's warning against regarding any such feature of the strategic scene as fixed and permanent.[55] From these shaky premises Bush inferred that 'For the near future [undefined], the really important and significant field of guided missiles lies in much shorter ranges . . . '.[56]

Those whom Bush was trying to refute, however, had seldom predicted ICBMs for the *near* future. From the point of view of the present enquiry, there is also another important aspect to the question of how feasible long-range ballistic missiles *could* have appeared to scientists in the 1940s. After the sputniks it was often argued that until the Operation Castle test explosions of potential thermonuclear warheads in April–May 1954 it was reasonable not to move the first ICBM project (Atlas) from paper studies into the initial phases of engineering development, because it was not certain before that point that the low level of accuracy achievable over intercontinental distances with forseeable guidance systems could be compensated for with warheads of far greater power than the physically-limited atomic bomb.[57] That argument neatly exempted the Truman administration from responsibility for the late American start on developing large rocket motors, and hence by implication from any share in the responsibility for the sputniks débâcle, and has been consistently popular with space historians ever since. However, it overlooks two inconvenient points. The first is that an initial capability for space exploration was not dependent on acquiring rocket engines of ICBM proportions, even if that was the way the Soviet Union chose to go about it. Something far smaller would have sufficed. The second is that the US Atomic Energy Commission had been working very effectively since 1947 to improve the ratio of explosive yield to weight in atomic warheads. If work on ballistic missiles had been linked intelligently to this programme, intermediate-range missiles like Thor and Jupiter would have been available two or three years sooner than they were.[58] Their rocket engines would then have provided the United States with a reliable satellite-launching capability during the IGY, rather than just after it, as actually happened.

# 4 Advice on Space for Truman

## 1 PUBLIC AWARENESS OF SPACE

In the late 1940s the American public was far more 'space-minded' than its government. The period between the two world wars had been a golden age of science fiction. At first most of the output in this genre had been almost too fanciful to merit the designation. In the 1930s the influence of the first serious technical studies in astronautics began to be felt, and the pulp magazines were provided with a modest but influential leaven of science-based speculation on the potential of rocketry for applications in space. From early 1947 public interest in such matters was further stimulated by a spate of reports of Unidentified Flying Objects (UFOs) which continued into the mid-1950s, giving rise to several official investigations.

One residual confusion, or oversight, affected most people who had come across the subject by the late 1940s. As terms like 'space flight', 'space ship', 'satellite vehicle', and 'astronautics' (coined in 1927) show, it was widely assumed that future space technology would be aimed primarily at conveying human beings into those regions beyond the Earth's atmosphere which have since become known as 'outer' space. People were slow to recognize that the first stage of astronautics would involve putting *unmanned* objects into Earth orbit to perform various scientific and technical tasks. The reason for this blind spot was that it took some time for non-specialists to appreciate the progress that was being achieved with the necessary miniaturization, automation and telemetry of potential spacecraft components, despite being told about it by journalists like the Alsop brothers or *The New Yorker*'s Daniel Lang.[1] Remarkably, one of the first references to possible uses of space technology in a non-specialist publication did in fact draw attention to a sort of unmanned satellite though without using the term. This was the playlet 'Pilot Lights of the Apocalypse' by Louis Ridenour, published in January 1946, in which one character remarked that:

> We have got pretty good information that some other countries . . . have got bombs up above the stratosphere, 800 miles above the earth, going round us in orbits like little moons.[2]

Ridenour knew what he was talking about. As one of his country's foremost authorities on radar, he had served as an adviser with the Army Air Forces during the war; later, when the US Air Force was established, he became its first chief scientist. By late 1945 he would have read von Braun's report on German wartime rocketry and its potential future applications, which had been written that summer while von Braun was being interrogated by Allied officers at Garmisch-Partenkirchen.[3] Shortly after 'Pilot Lights' Ridenour wrote the keynote chapter for the first report from Project RAND on potential applications of satellites (see section 3 below).[4]

After the classification on the von Braun report was relaxed in January 1947 it became available as an important source for the article by the Alsops. In addition to long-range missiles they looked at the possible future applications of an 'unpiloted satellite' and quoted directly from von Braun's text. (This was the first time, apart from a few lightweight interviews with the El Paso *Herald Post*, that any of von Braun's ideas were presented to the American public in his own words.) However, the quotation was edited to conceal the fact that in the original document von Braun had been discussing the 'construction of multistage *piloted* rockets' and had neither limited himself to, nor even shown much interest in, unmanned applications.[5]


## 2   GEOPHYSICS AND WARFARE

In the aftermath of the development of the atomic bomb by the Manhattan Project and the post-war revelations of German achievements in aeronautics and many other fields, the Truman administration was occasionally told that military power in general, and air power in particular, would in future 'be built around scientists', as General Arnold expressed it.[6] But amongst 'the remaining secrets of the universe' which the Alsop brothers declared that 'it will first be necessary to ferret out' in order to build long-range missiles, three areas of ignorance were particularly relevant to the development of a space programme.[7] They were those in which satellites carrying automatic measuring instruments would eventually provide some of the scientific prerequisites for operational ICBMs. (This is a reversal of the dependence relationship which is usually thought to link the two military applications of rocketry, simply because large propulsion systems were first developed for long-range missiles and then used in satellite-launchers.) The scientific fields in question dealt first with the density, composition and vertical extent of the upper atmosphere, including variations caused by such factors as the seasons or irregular solar activity; second, with the nature and composition of the ionospheric layers; and third, with the precise 'figure of the Earth' and the structure of its gravitational and magnetic fields.

## Atmospheric Sciences

In the second half of the 1940s the Earth's atmosphere was thought of as composed of four vertically adjacent regions, each of which gradually merged into the next. Below about 45 000 feet was the troposphere, containing three-quarters of the atmospheric matter, including almost all the water vapour and dust, and consequently almost all the 'weather'. Above this, up to about 340 000 feet, was the stratosphere, containing just under a quarter of the total matter. Then came the ionosphere, with about 1/3000th part of the total, in which several poorly understood radio-reflecting layers were known to exist. Above this, from about 1 900 000 feet or 350 miles, the remaining minute traces of the atmosphere, about one hundred-billionth part of the whole, were thought to merge gradually into the void of outer space. This outermost region was termed the 'exosphere'.[8]

The maximum height that had been reached with uninstrumented sounding balloons was 120 000 feet, well into the stratosphere but only a slight improvement over pre-war figures. The maximum for a manned aircraft remained the 72 395 feet, slightly less than 14 miles, attained by Orvil Anderson and Albert Stevens with the balloon Explorer II in November 1935. The apogee of the normal trajectory of wartime V-2s had been between 50 and 56 miles, or about 280 000 feet, though a vertical altitude record of 117 miles was also achieved during wartime testing.[9] The design of aircraft and missiles to fly within or to re-enter from outside the upper atmosphere obviously called for greater knowledge of the region. When Convair's MX-774 ballistic missile study was cancelled in 1947, one reason mentioned by the head of the Air Materiel Command was that studies to overcome 'the very serious temperature difficulties on re-entry' were bound to be long and costly 'inasmuch as so little is presently known about the properties of the upper air'.[10]

Even before the war ended there were proposals for the V-2 rocket to be adapted for upper atmosphere research, or for the development of specialized meteorological rockets for the same purpose.[11] One such scheme was based on comments supplied by the acting head of the US Weather Bureau.[12] In February 1946 representatives from the armed services and several universities came together under the auspices of the Army Ordnance Corps to form the V-2 Upper Atmosphere Research Panel. The panel's initiator and first chairman was Ernst Krause from the Naval Research Laboratory; he was succeeded in December 1947 by James Van Allen from the Applied Physics Laboratory (APL) at Johns

Hopkins University. The first captured V-2 carrying a full instrument payload for atmospheric measurements was launched to an altitude of 67 miles from the White Sands Proving Ground (WSPG) on 28 June 1946, over a year before any comparable launch in the Soviet Union.

The programme continued until 1952. After 35 missiles had been assembled from captured parts, further rounds often required new parts to be supplied from the WSPG workshops. At first modifications and then complete new sounding rockets were introduced by American engineers. In March 1948 the panel was renamed the Upper Atmosphere Rocket Research Panel.

## Guidance

Global communications in 1945 were based on a network of overland telephone and telegraph cables, underwater telegraph cables (some trans-oceanic), and high-frequency (HF) radio links. Because of the distances they could cover and their relatively high channel capacity, first demonstrated by the Marconi company in the 1920s,[13] HF radio links were considered vital for rapid communications in times of crisis or war. But radio only fulfilled this function by means of the reflective properties of little understood layers within the ionosphere. Ionospheric disturbances could and frequently did lead to widespread communications failures.

The relevance of these facts to the prospects for long-range missiles lay in the widely canvassed suggestion that their accuracies could be improved by transmitting in-flight guidance commands, along lines followed in several short-range types. It was obvious that cruise missiles could, in theory, be controlled in this way. As for long-range ballistic missiles, they might perhaps be designed as winged 'glide bombs', which would not only reduce the heat and other stresses of atmospheric re-entry but also provide aerodynamic control surfaces for course corrections during the approach to the target, or 'terminal phase'. American experts at first adopted a version of this idea from von Braun's 1945 survey of the German missile programme, in which the bombing missiles would be guided by signals from the crew of a 'piloted rocket', or space station, 'equipped with very powerful telescopes'. The offensive missiles might also be stored on and launched from such 'true space ships'.[14] The alternative of using an unmanned satellite for the same purpose was also endorsed at a meeting of the Research and Development Committee of the War Department's Aeronautical Board in May 1946.[15] At other times the proposal was to use control signals from aircraft at high altitude, but the implications for ionospheric physics were the same.

## Targeting

The inadequacy of available maps for supporting automatic navigation on long-range bombing missions had already been experienced in World War 2. Less than half the world's surface had been mapped to the level of accuracy then attainable. For post-war American operations planning this problem was exacerbated in the case of targets inside the Soviet Union:

> The enemy's territory was vast and unknown. Target planners had to rely on pre-World War II and even Tsarist era maps, or at best German aerial photos from 1942–1943.[16]

Before the first satellites were launched it was generally assumed that intra-continental mapping in Europe and North America was fairly precise, and that only the relationship between the two systems was affected by major uncertainty, of the order of 100 parts per million, which could lead to an error as great as 2000 feet in calculating the distance between New York and Paris. In the 1950s, however, it was discovered that maps of Panama had failed to include a 125-mile long range of mountains, with peaks as high as 5000 feet.[17] Tests of missile navigation systems sometimes revealed serious mapping errors even within the United States. For example, work on the Navaho guidance system showed that 'El Paso is a mile and a half away from where it is supposed to be'.[18] Major improvements in geodesy were obviously needed before long-range weapons could be regarded as feasible.

\*    \*    \*

The importance of these three areas of scientific research and technological development for long-range ballistic missiles is certain to have been emphasized in a secret report on *Geophysical Sciences and Warfare*, written for the RDB's Geophysical Sciences Committee in 1948. Apart from radio communications for in-flight guidance of long-range missiles, all the relevant points were made in the sanitized version released a few years later.[19] (They were repeated during the Eisenhower administration, usually in guarded terms or in off-the-record asides, by American scientists seeking Congressional funding for their activities within the International Geophysical Year.)[20]

## 3  THE NAVY AND RAND SATELLITE STUDIES

Hindsighted recollections of the prescient space-mindedness of this or that group or individual in the United States have been plentiful since 1957. One of the stranger ones came in a lecture given by Lloyd Berkner in January 1962. Berkner claimed that the Lincoln Summer Study had 'advocated that the United States seriously undertake a space program' but that 'the idea was hooted down as outrageous'.[21] This appears to be the only such description of the Summer Study on public record.[22] It may be accurate, but that cannot be verified until the Study's report is declassified.

In the immediate post-war years the term 'satellite' was slow to lose its connotations, derived from planetary astronomy, of a relatively large and therefore conceivably habitable object. It did so most rapidly within a series of studies conducted by Douglas Aircraft's Project RAND, which became the RAND Corporation in November 1948 (below). Elsewhere people continued to feel a need for special expressions such as 'orbital rocket' or, at best, 'satellite rocket', to denote the unmanned variety of spacecraft. Confusion could therefore arise when, for example, the 1946 RAND report referred to an unmanned object as a 'spaceship'.

Berkner's statement is interesting because it suggests that *before* 1952 he had shared the negative attitude to the early post-war satellite proposals that was adopted by Vannevar Bush and the RDB. However Captain W. P. Cogswell, who worked alongside Berkner at the Navy Bureau of Aeronautics (BuAer) when the earliest post-war satellite studies were being carried out in 1945–6, has given a very different account of Berkner's thinking at that time.[23]

In 1945 Captain Lloyd Berkner (USNR) was Head of the Electronics Materiel Branch of the Engineering Division at BuAer. Attached to the branch was Lieutenant Robert Haviland, one of whose principal duties it was to examine Allied intelligence reports on aviation matters. In this capacity Haviland read a British report on the V-2 rocket. He also became aware of discussions of possible applications of rocketry which had begun in the Bureau's Aviation Design Research Branch in July, when a copy of von Braun's report on the German rocket programme was brought in by the physicist Abraham Hyatt, who had been in Europe earlier that year with the Navy's Technical Mission. Haviland had been interested in the possibility of space travel for some time, and he now combined his new information with ideas taken from Willy Ley's *Rockets* and from other sources, to produce a nine-page memorandum to the head of BuAer's Special Weapons Section, Commander J. A. Chambers, recommending that the Navy undertake an Earth satellite project.[24] One factor which caused Haviland's superiors to notice his proposal was its proximity to President Truman's announcement on 6 August that the United States had developed and used the atomic bomb.

According to Cogswell, Berkner was particularly taken with the idea that satellites could serve as 'an artificial ionosphere', or in other words as (passive or active) radio repeaters.[25] Berkner in turn convinced his superiors that it would be worth investigating whether satellites were possible and what they might be used for, and brought Commander Harvey Hall, a senior BuAer scientist, into the process. Hall formed the Committee for Evaluating the Feasibility of Space Rocketry (CEFSR) on 3 October 1945. As more experienced scientists became involved Haviland's visionary ideas about manned spacecraft were abandoned and a proposal was adopted in late October 1945 for a single-stage unmanned satellite, or 'High Altitude Test Vehicle'. This depended on developing a fuel with a much higher specific impulse than anything then available, but CEFSR members were encouraged in that respect by the ongoing achievements in rocket propulsion technology at the Aerojet-General Corporation, under Navy contracts.[26]

In December 1945 BuAer commissioned an independent report on the CEFSR findings from the Jet Propulsion Laboratory. It was completed in July 1946 by Homer Stewart, who determined that the original CEFSR design could not achieve an orbital insertion higher than 50 miles, at which altitude air resistance would prevent the achievement of even one complete orbit. Stewart suggested that such problems would be far more tractable with a two-stage rocket, but Hall and BuAer remained wedded to the single-stage approach, for which they put out preliminary design contracts to North American Aviation and to the Glenn Martin Company in association with Aerojet, as well as conducting an in-house design study of their own under Hyatt.

By early 1946 BuAer had decided that a full-scale engineering design project would cost more than they could hope to obtain in the context of the budget cutbacks then being required of the armed forces.[27] In March they approached the Army Air Forces with a view to a joint project. Major General Curtis LeMay, the newly appointed Deputy Assistant Chief of Air Staff for R&D, consulted the head of the AAF, General Spaatz, and then told the Navy their proposition could not be accepted.

The AAF had already begun to make their own plans for a role in outer space. General Arnold's report in November 1945 was followed in December by a proposal for a new engineering development centre which should have amongst its tasks the development of 'flight and survival equipment for use above the atmosphere, including space vehicles, space bases, and persuasive devices [*sic*] for use therein'.[28] While stalling Navy requests for further meetings, the AAF sent an urgent request to Douglas Aircraft's newly formed Project RAND for their own satellite study. Sixteen experts laboured to produce an impressive 250-page report by 2 May 1946,[29] which enabled General Laurence Craigie to adopt the 'bargaining position that the AAF was on an equal or similar development position with the Navy' when the two sides met at the Research and Development Committee of the War Department's Aeronautical Board thirteen days later. The Navy stated its position in favour of a joint project; the Army replied 'that because intercontinental warfare is of paramount interest to the Army Air Forces, future plans involving actual construction of an earth satellite should be under the control of that organization'. 'From this point on,' recalled Harvey Hall, 'no further progress was made towards a joint project.'[30]

Interest in the satellite idea at BuAer may have been revived by this rebuff, and design studies progressed satisfactorily through to the end of a Fiscal Year in June 1947. In January 1947, however, Rear Admiral Leslie Stevens, BuAer's Assistant Chief for R&D, wrote to the JRDB suggesting that coordination of interservice requirements for Earth satellites lay beyond the scope of the Aeronautical Board, and should now be carried out by an *ad hoc* panel of military and civilian experts under the JRDB. This was essentially a repetition of the former Navy proposal for a joint project, intended to prevent an Air Force monopoly.

Eventually the Research and Development Board in the new Department of Defense accepted the negative part of Admiral Stevens's proposal, and removed the satellite issue from the competence of the Aeronautical Board. Instead of setting up a new agency to handle it, however, they assigned it to their own Committee on Guided Missiles. Meanwhile BuAer officials, faced with revised cost estimates of $50 to $150 million for placing a prototype satellite in orbit, took no new initiatives for an autonomous satellite programme. In March 1948 a Technical Evaluation Group established by the RDB's Committee on Guided Missiles reported that a satellite was technically feasible, but found adversely on the all-important question of its near-term utility at the estimated price. This meant an end to the Navy's single-stage approach to a satellite project, and when the Navy failed in its attempts to become a co-sponsor of RAND alongside the Air Force, to the end of the Navy's involvement with the idea. All activity arising from the early Navy satellite studies had ceased by the end of 1948.

In February 1947 Project RAND produced an improved design for a satellite system in a collection of reports which noted the need for further research in the fields of guidance, on-board power systems, attitude control, and telemetry. (At the time, such technologies as solar batteries, electronic miniaturization, and computers were still in their rude infancy, by comparison with what they were to become even by the mid-1950s.) Then, as has been recalled by Robert Salter, who subsequently directed RAND's Project Feed Back on reconnaissance satellite technologies:

> There was a period of time between February '47 and the very latter part of '49 when the amount of work done on the Satellite program was virtually nil. During this period of time jurisdictional decisions were being made and people would generate some documents to support their particular case with the RDB but nothing active was formed along this line. . . .
> . . . in the year 1948 . . . nothing was going on in the satellite program at all.[31]

The problem seems to have been a lack of enthusiasm for potential space technologies on the part of RAND's sponsor, the US Air Force, especially at the Air Materiel Command. However, in 1949 and 1950 RAND staff resumed work on the twin tasks of evaluating technical progress resulting from American rocket programmes, and of re-assessing satellite utility. Several reports were produced between 1949 and 1951, of which those dealing with satellite reconnaissance and related applications remain classified today.[32] One of the participants has recalled that Project Feed Back, which completed its final report in 1954, was in some ways a last-ditch attempt to convince a reluctant Air Force of the value of the satellite idea:

> The RAND scientists were now beginning to become impatient and frustrated. First they demonstrated feasibility, then utility; still there was not enough support within the Air Force or the Defense Department to start development. RAND was to make one more study, called Project FEED BACK. This project was to design an observation satellite with sufficient detail to prepare a development plan.[33]

An important aspect of Feed Back was the participation of industrial companies such as RCA, Westinghouse, and North American Aviation.

By 1953, designs for all elements in a satellite reconnaissance system
were at an advanced stage, but after seven years of paper studies not a
single piece of development work had been commissioned.

## 4   OTHER EXPERT ADVICE ON SATELLITES

Few public discussions of space technology were concerned with the
possibilities of *unmanned* satellites before 1949. One exceptional article,
published in *Army Ordnance* at the end of 1946, mentioned this aspect
of the matter first, before passing on to discuss the possibilities of
manned space stations and long-range missiles. The pseudonymous author
explained why and how such 'gravity-free rockets' would be a useful
extension of the research already being done with instruments carried aloft
on V-2s and balloons.[34]

This point was also made quite widely before 1949 in lectures and
interviews given by scientists involved in the V-2 programme, such as
Charles Green, a consultant engineer with General Electric and member of the V-2 Upper Atmosphere Research Panel.[35] But sometimes
anxiety about the possible public reaction kept people from going too
firmly on the record. James Van Allen included a closing paragraph
about satellite possibilities in a paper sent to an international scientific conference in Oslo in August 1948, only to find his ideas greeted
with levity by *The New Yorker* and with cynicism by *The New York
Times*.[36] On the advice of the director of his laboratory (APL) he reluctantly removed the passage from the published version, although he
has since remarked that 'in my professional circles, I was not making
a very revolutionary suggestion at all. It wasn't original with me. It
was part of the general thinking and kind of society I was associated
with.'[37]

Four months later American journalists began to reconsider the scoffing
approach they had generally preferred for items about space and satellites
before then. The new Secretary of Defense James Forrestal published
a *First Report* containing the statement that the armed services were
conducting separate but (allegedly) coordinated studies of such a possibility.[38] (The announcement itself will be further discussed in Chapter
6.) The effect of this development on public perception is evident from
headlines such as 'U.S. Military Men Really Plan Earth Satellite Vehicle'.[39] *Life* ran a story on space flight which included a misleading
claim that the Moon was being considered by 'forward-looking military leaders . . . as a base from which rockets could be launched to

control the world'. But it was also stated that the immediate proposal was simply to place in orbit 'an unmanned spaceship' that could 'automatically radio back information' of possible use for future space projects.[40]

There followed the launching of a two-stage sounding rocket (Bumper-WAC) to a record altitude of almost 250 miles, at White Sands in February 1949, and the space exploration lobby began to get a better reception. Willy Ley, for example, published articles on rockets and satellites in *Scientific American* and *Technology Review*, as well as a popular book on space travel illustrated by Chesley Bonestell, who was to become America's leading space artist.[41]

Members of the British Interplanetary Society played an important role in developing the discussion of unmanned satellites in the open literature. Kenneth Gatland's first commentary on the American announcement still showed the traditional fixation on manned space stations, with its talk of a full-scale observatory or a variety of research laboratories in orbit.[42] In the next two years, however, Gatland and two colleagues published articles about applications of unmanned satellites for the initial phases of a space programme. In the second of these, which was particularly influential, the authors examined the feasibility of using a 'step-rocket' to launch either an inert 154 lb satellite, or instrumented satellites with total masses of 772 lb, 1235 lb or 1962 lb, and with 220 lb, 220 lb, and 485 lb payloads respectively.[43]

To put those numbers in perspective, in December 1949 Ley had discussed a project to orbit a satellite with a total mass of 1000 lb and a payload of 200 lb. The subsequent American project for an IGY satellite began with hopes of orbiting payloads of 30–50 lb, but eventually reduced the amount to 21.5 lb. The first two actual American satellites, in 1958, carried 18 lb and 3 lb of instruments out of total masses of 31 lb and 53 lb in orbit. Obviously the space enthusiasts of the early 1950s were still some distance from a realistic appreciation of what might be feasible in the near future with Western rocketry.[44] (The first two sputnik launches each placed about 9000 lb gross weight in orbit, with payloads of 184 lb and 1120 lb respectively.)

Even after the so-called 'Forrestal announcement', or perhaps because of the mixed reaction to it, the gulf between what scientists and engineers were talking about amongst themselves and what they were prepared to say in public was slow to disappear. In February 1950 Homer Newell, an atmospheric physicist at the Naval Research Laboratory (NRL), wrote merely that all upper atmosphere rocket research to that date had been just a beginning, and that 'The ultimate potentialities of the rocket, and perhaps some day the satellite vehicle, are yet to be realized.'[45]

Outside the specialized journals of the rocket societies there were few public discussions of astronautics in 1950. One landmark was a short technical study of space flight by Arthur C. Clarke, then chairman of the British Interplanetary Society.[46] Another was the foundation of the International Astronautical Federation (IAF) at a meeting in Paris in October. The only delegate from outside Europe came from Argentina, though one American student in Paris, Frederick Ordway, attended the public session as an observer.

*                    *                    *

In 1951 there were two important developments for the public discussion of astronautics in the United States. The first was an invitation to Willy Ley to organize a symposium on space flight at the Hayden Planetarium in New York. The second was a decision somewhere in the US Army to let Wernher von Braun off the leash. Ley's symposium was held on 12 October, Columbus Day, before an invited audience of about 700 people. The speakers were Ley himself, Robert Haviland, Frederick Whipple (the Harvard astronomer) on relevant aspects of the space environment such as meteorites, Oscar Schachter (deputy director of the UN's legal department) on space law, and Heinz Haber (a founding member of the Department of Space Medicine in the USAF's School of Aviation Medicine) on the rapidly expanding field of space medicine.

The symposium received considerable publicity, and was followed by an invitation from *Collier's* magazine to publish an illustrated version of the contributors' ideas. However, the original arrangement had been less than satisfactory because Haviland, who had been invited to present the key paper on 'The Engineering and Application of the Space Vehicle' (i.e. of an orbital space station), was barred from doing so at the last minute by his employers, General Electric, and had to fall back on a mundane account of the Bumper-WAC launchings instead.[47] Fortunately for *Collier's* the Army had relaxed its previous restrictions on von Braun's public activities by this point, so that he could contribute the necessary article instead of Haviland. With the support of Bonestell's magnificent illustrations it dominated that issue of *Collier's* and, in an amended version, the subsequent book.[48]

Up until 1951 von Braun's position was a delicate one. He had entered the United States in September 1945 under the Army's covert Operation Paperclip, and was not granted citizenship until nearly ten years later. As details of Paperclip became known, the status of former Nazi German scientists in the United States formed a focus of public controversy.[49] In this context, a request from the American Rocket Society to the Army Ordnance Corps, for von Braun to feature as a luncheon speaker at their 1947 annual meeting, was predictably unsuccessful. For the same reason, although von Braun sent papers to the IAF Congresses in 1951, 1952 and 1954 he was unable to attend and deliver them in person. In March 1950 and November 1951, however, he took part in public symposiums on space medicine in Chicago and San Antonio. His contributions to the open literature of astronautics began in 1951, when a paper to the first space medicine symposium and a digest of that sent to the Second IAF Congress were printed.[50]

According to the biographers of 'the rocket team' von Braun made a conscious decision in about 1949 or 1950 that it was time to 'talk to the people'.[51] In 1952 he really got going. The special *Collier's* issue was launched with a lecture on 19 March at the White Oak Naval Ordnance Laboratory, near Washington.[52] The hall was crammed with an audience of about 2000 people and several invited dignitaries were unable to get in.

In October 1952 *Collier's* followed up its earlier success with a sym-posium on lunar exploration, to which von Braun, Ley and Whipple contributed, with Bonestell and others illustrating. Once again von Braun's article was linked to a public lecture – this time it was his first one to a Hayden Planetarium meeting – and the material was expanded into a book.[53] Although he accepted the need for 'development of a smaller, unmanned orbital rocket' before 'the manned satellite ship', his manner of doing so was characteristic. In his view, the unmanned vehicle should be integrated within a 'well-planned strategy' aimed pri-marily at achieving the space station, which meant designing the first and second stages of the unmanned satellite-launcher so that they could also serve as second and third stages for the larger rocket. He therefore proposed that the upper stage of an unmanned orbital rocket should be 'of the dimensions of the V-2 or the "Viking"', so that it could later be adapted, at a pinch, for a human passenger.[54] The impractical scale of his ideas was also revealed in his sole publication on the unmanned satellite, a *Collier's* feature in June 1953 entitled, typically, 'The Baby Space Station'. By that point, he claimed, 'Scientists and engineers working toward man's exploration of the great new frontier *know now* that they are going to send aloft a robot laboratory *as the first step*.' This allegedly 'miniature' device was to be a 30-foot long object weighing several tons and carrying three rhesus monkeys 'in air-conditioned comfort'.[55]

Von Braun did not have things all his own way. At the second Hayden Planetarium meeting in October 1952 his large-scale plans for human space travel were criticized as impractical and misleading by Milton Rosen, director of the Naval Research Laboratory's Viking sounding rocket programme. In Rosen's opinion, the von Braun space station concept was 'fantastic' and rested on assumptions that were 'pure and unfounded speculation'.[56] This famous and to some extent deliberately staged confrontation made the cover story in *Time*.[57] The debate between the partisans of manned and those of automated spacecraft has continued ever since.

*         *         *

Although it was not strictly relevant to the debate about the advisability of proceeding with von Braun's 10-year, $4 billion programme to put a manned space station in orbit, the question of the possible military applications of such an object undoubtedly influenced the judgements of most contributors to the discussion. Von Braun lost few opportunities to mention his long-held idea about arming a space station with atomic weapons so that 'it would represent an ever-present guardian against war and the usurpers of power would constantly be reminded [by its visibility] that they would sow world discord at their own peril'.[58] Rosen, by contrast, 'object[ed] strenuously to the contention . . . that either we must build a space station to dominate our enemies or they will build one to dominate us'.[59]

Others criticized the idea on military grounds, arguing that an orbital bombing platform would be too vulnerable to attack to constitute the sort of absolute weapon that von Braun appeared to believe it.[60] While the experts argued, the image of an atomic sword of Damocles in Earth orbit, first aired by Ridenour over six years earlier, became a powerful and enduring one in American popular culture. It can be traced at the outset of the Eisenhower administration, for example, in William Menzies's successful film *Invaders from Mars.*

*         *         *

A major field of research which contributed to astronautical ideas after World War 2 was that of space medicine. This discipline was established in the United States largely by transferring knowledge and experts from Germany, first by holding a seminar at the Heidelberg Aeromedical Centre in 1946, and then by importing the director, Hubertus Strughold, and several other staff of the Aeromedical Research Institute of Berlin under Operation Paperclip, in 1947.[61] A panel discussion on the subject was presented to a largely military audience of 700 people in November 1948, and a Department of Space Medicine was set up as part of the USAF Air University's School of Aviation Medicine at San Antonio the following year, with Strughold as director.[62] In March 1950 the first conference on space medicine was held at the University of Illinois, and von Braun gave one of his first public presentations of the space station plan which he was soon to put before millions of Americans. It was only slightly more detailed than his report to Allied interrogators in 1945:

> Our space station could be utilized as a very effective bomb carrier, and for all present-day means of defense, a non-interceptible one. If we fire a small rocket missile . . . By means of remote control and radar observation we may guide such a missile accurately to any spot on the map.[63]

In November 1951 400 military and civilian scientists gathered in San Antonio for a conference on 'The Physics and Medicine of the Upper Atmosphere'. Alongside cosmetic phrases about 'very high' and 'upper' altitudes there were frequent references to 'manned rocket flight', 'satellite vehicles', and 'space ships'. As Robert Salter bluntly expressed it, 'we are to discuss the "how" and "when" of manned space flight'.[64] There were no papers from sceptics. Several journalists attended, and the meeting was widely reported. However, discussions of this kind and the associated publicity did little to focus attention on the actual possibilities and requirements of a preliminary space programme for the 1950s, which remained those for unmanned spacecraft.

*     *     *

In general, it is evident from this survey not only that sound advice about potential space technologies was available from several quarters during the Truman administration, but also that agencies within or close to the government, such as the Office of Naval Research or the RAND Corporation, had at least as clear a picture of the route that would need to be followed into space, and of the advantages to be gained by following it, as any outside bodies. The principal factor operating against their efforts to get a satellite project off the ground appears to have been a dogmatic conviction, above all at the top of the US Air Force, that the time was not yet ripe for such a venture. The interaction between that doctrine and intelligence assessments of Soviet research and development will be examined in Chapter 5.

# 5 Intelligence for Truman

## 1 THE SOVIET MISSILE PROGRAMME

Before evaluating the intelligence available to the Truman administration about post-war Soviet R&D for missiles and other applications of rocketry, it is useful to summarize what the Russians were actually doing at the time, in so far as that is now known to Western historians. During World War 2 the pre-war Soviet rocket development programme, which had been larger and more advanced than anything in the West outside Germany, was virtually suspended. At the end of the war the Soviet government immediately established a Scientific-Technical Council for Rocket Development, directed by People's Commissar Dmitri Ustinov.[1] Its most urgent task was seen to be the assimilation of the recent German technical achievements.

At the beginning of 1945 over 2000 of the 4325 employees at Peenemünde were still working on the A4 rocket (V-2) programme, and over 1200 on developing the new Wasserfall air-defence missile. When von Braun and his colleagues decided to accept an order to move the entire project several hundred miles south to the Nordhausen-Bleicheröde area of Thuringia (one of several conflicting instructions), almost a third of the staff stayed behind. Out of some 3000 who made this move, von Braun was then ordered to make a second selection of only 500 key personnel to be transported a further 400 miles south to the Bavarian Alps. As was related in Chapter 3, after surrendering to the US Army less than half of this élite group were accepted for transfer to the United States under Operation Paperclip. A small number of others found refuge and employment in France and Britain. There were thus several thousand people, some fairly senior, who remained available for recruitment into the Soviet post-war missile programme. In 1945 many still lived either near Peenemünde, or in Thuringia, both districts which were included in the Soviet occupation zone.

The Zentralwerke organization for rocket development, with 5000 employees, was promptly established in the Soviet zone under Helmut Gröttrup, the former assistant director of the Guidance, Control and Telemetry Laboratory at Peenemünde. Despite the American capture of parts and papers it took only a year to replace the blueprints and start a pilot production line delivering flightworthy V-2s. In October 1946 200

Zentralwerke staff, headed by Gröttrup, were suddenly taken to the Soviet Union, together with other experts from related institutions. One year later the first Soviet V-2 flight-tests began, and a series of about twenty shots was fired in five weeks. As in the United States, several carried instruments for upper atmosphere research. After working on the design of several missiles with ranges of between 75 and 1800 miles, none of which was ever built, all but twelve of the German rocket experts were returned to East Germany in 1951 and 1952, whence several promptly made their way to the West.[2]

Because of the legacy of indigenous expertise from the pioneering work done in the 1930s, the post-war Soviet organization for rocket development was never dependent on the imported Germans, who were allocated largely supporting roles such as checking on design calculations.[3] High priority was given to the establishment of a nationwide complex of rocket research facilities, design bureaux and factories between 1946 and 1947, and in March 1947 Stalin personally intervened to establish a new state commission for rocketry.[4] The first rocket units were formed within the Soviet Army as early as 1946.[5] Their initial armament was probably the short-range (20-mile) unguided artillery rocket known as 'Frog-A' in the West. This was followed by the Soviet Union's first guided ballistic missile, the conventionally-armed 'Scud-A' (50-93 miles), by copies of the V-2, and in about 1950 by a nuclear-capable 'improved' V-2, the Pobeda, with a range of 560 miles.[6] The first division of a new arm of the Soviet military, the Strategic Rocket Forces, was formed in 1950 with V-2s and Pobedas. By 1956 the Pobeda's range was being extended to 700-750 miles.[7] Although the United States never developed an exact equivalent in the interval between its 200-mile Redstone (operational from 1958) and its later IRBMs, the Pobeda can be seen as placing the Soviet Union two or three years ahead in the missile race.

## 2 EARLY ESTIMATES OF SOVIET ROCKETRY

Clearly there was a major Soviet effort in rocketry in the 1940s and early 1950s about which the Truman administration would have preferred to be fully informed. But in those days the plethora of technical intelligence-gathering systems which have dominated the post-war espionage scene still lay in the future. American intelligence agencies lacked, not merely satellite-borne surveillance systems, but also some of the larger radio stations for the collection of signals intelligence which were to be built in the 1950s and 1960s. In particular, the American listening-stations in Turkey, which were able to monitor test-flights from the first large Soviet missile development centre at Kapustin Yar, were not operational until 1955. Less specialized listening-stations had of course been used in World War 2 and continued in operation against new targets. Amongst them were the American Kagnew station in Eritrea and British stations at Gibraltar and at Habbaniyah in Iraq. New stations were rapidly acquired in West Germany, Austria and Iran, and in 1947 the UK-USA Security Agreement was signed, dividing the responsibility for collecting signals intelligence around the world between the United States, Britain, Australia, Canada, and New Zealand.[8] These passive forms of technical intelligence-gathering were supplemented with more active ones, as American and British planes crammed with electronic and photographic equipment flew along and sometimes for short distances inside the borders of the Soviet Union and its allies. Truman did not, however, authorize an extensive programme of shallow overflights of the Soviet Union until 1950. Kapustin Yar lay beyond their normal penetration range, and seems not to have been directly overflown until 1953, after Truman had left office, by a British-manned B-57 which was intercepted and nearly brought down by Soviet fighters.[9]

Western intelligence agencies were therefore heavily dependent on more traditional sources of information, such as published material, spies and defectors, during the Truman administration. These were by no means lacking. Before Stalin's personal initiative moved the Soviet rocket programme into top gear in 1947 articles showing military interest in the potential of long-range rocketry occasionally appeared in the open Soviet press. In May 1946, for example, General M. Gerasimov referred to 'the significance of rocket artillery, which is difficult to detect but capable of firing projectiles with sufficient accuracy to destroy targets located hundreds and possibly thousands of kilometers away'.[10]

Several books on rocketry were published in the Soviet Union in the 1940s. They included a new edition of the works of the great Russian pioneer of astronautical theory, Konstantin Tsiolkovsky, in 1947, a study of the *Chemistry of Rocket Fuels* in 1948, and an account of *Russian Rocket Artillery* in 1949.[11] The extensive Russian speculative literature on rocketry and spaceflight from the pre-war period also remained in circulation. Rear Admiral Leslie Stevens, the former US Navy R&D chief who had supported early satellite studies in the United States, personally encountered an example of this material in Moscow in the late 1940s, and later recalled receiving the impression that the Russians were likely to pioneer in the field of space exploration.[12]

*             *             *

Early post-war American intelligence estimates of the relative achievements in rocketry R&D of the United States and the Soviet Union tended to be cautious, or even pessimistic. This was probably related to a wider appreciation of the extent to which the United States had cut back on advanced military R&D during the war. Thus in July 1946 the US Strategic Bombing Survey remarked that:

> It is not generally realized the degree to which basic scientific research was neglected in the United States during the war . . . nor the degree to which we lagged behind Germany in advanced aerodynamics, jet propulsion, and the development of guided missiles. In air armament and torpedoes, even the Japanese were ahead of us. One or two years' lag in either basic research or in the development of reliable military application of such research can only be made up with difficulty, if at all.[13]

In March 1947 General Curtis LeMay, the Deputy Assistant Chief of Air Staff for R&D, testified that:

> . . . the United States is far behind in certain fields of research and development, particularly in the sciences and techniques associated with guided missiles. I would like to call your attention to the reports recently in the newspapers of missiles flying over Sweden and Greece. We definitely are a year or more behind in some phases of jet power plant development.[14]

By contrast, in October 1947 Rear Admiral Daniel Gallery, who was one of the more missile- and even space-minded US Navy chiefs of the day, told John McCone, then serving on the President's Air Policy Commission under Thomas Finletter, that, while it would take the United States five years to develop an effective anti-aircraft rocket, it would take their 'potential enemies' twice as long, because:

> In regard to Russian capabilities, I think they are definitely below ours. I realize that such a statement is no help in building up a case for larger appropriations, and maybe it is rather smug to make it. But the Russians have to rebuild a large part of their industrial plant, which never was anywhere near the equal of ours, and I don't think they have the talent, facilities or resources to put on the job that we have.[15]

In Gallery's view the supersonic missile for intercontinental bombardment was about ten to fifteen years away. McCone passed this estimate to his aides, Colonel Bryant Boatner and Captain Paul Pihl, who then wrote a memorandum for the Air Policy Commission which increased the interval to between fifteen and twenty years. However, they added a word of caution on the all-important question of who was likely to get there first:

> If the Russian acquisition [of German rocket technology] is well rounded, it is possible that a major portion of the German advance will be transferred to the Russian program. On this basis, Russia could now be further advanced in guided missile development than is the United States.[16]

Boatner and Pihl were conscious of an important fact which seems to have passed Gallery by, namely that unlike the United States the Soviet Union had had a major missile programme since the early 1930s. A contemporary intelligence report to the Joint Chiefs of Staff even concluded that the United States would have to spend many years *catching up* with the Soviet Union:

> Our enemies will have weapons generally comparable to our own, but we have the capability of attaining [parity by 1955] and retaining superiority of performance.[17]

In October 1948 an Army intelligence report declared that post-war Soviet work on rocket development, assisted by imported German experience, was definitely superior to its American equivalent.[18] And at the end of 1949 a CIA report on Soviet combustion research for jet propulsion and rockets found that Soviet scientific achievements in this area 'are regarded highly by U.S. scientists' and that a past 'shortage of engineering "know-how"' in the Soviet Union was 'rapidly being bridged'. With what is now known to have been total accuracy, the CIA analysts continued:

> It has been reported that the Soviets have definite plans for producing a new type rocket of their own, rather than German, design. We have no information to indicate that German scientists are working on any of the theoretical or fundamental experimental work referred to in this report.[19]

## 3   THE DEFECTORS

Important intelligence about the Soviet guided missiles programme was brought to the West in 1948, through the defection of Lieutenant Colonel

Grigori Tokaev, a former head of the Aerodynamics Laboratory in the Zhukovsky Academy of Aeronautics in Moscow. Tokaev had played a leading part in Soviet post-war technical intelligence operations in Germany, and then served as chief scientist on Stalin's special state commission. He testified at first hand as to Stalin's interest, particularly in the development of intercontinental weapons. Ironically, his credible if one-sided account portrayed Soviet leaders as agreeing with his own assessment, in the late 1940s, that 'we were far behind other nations in the sphere of reactive and rocket technology'. This finding was perhaps related to his belief that Texas had already been transformed into 'a vast Peenemünde', which was a considerable exaggeration, possibly arising from its author's ignorance of matters American.[20]

In its original version, Tokaev's story was that Soviet aspirations for an intercontinental 'missile' had focused on the aerodynamic skip-bomber concept devised by Eugen Sänger and Irene Bredt, which had been placed before the Nazi leadership in August 1944.[21] Their idea was that a rocket-powered manned bomber could be boosted into space to an altitude of about 250 miles, and could then descend to 'ricochet' several times along the top of the denser portion of the atmosphere. Such a rocket plane could achieve intercontinental ranges in excess of 7000 miles without needing to carry large amounts of fuel or to be launched with a massive rocket booster.

Tokaev made much of a conversation he had had with Malenkov in early 1947, in which the latter had complained of the inadequacy of the 'V-2s' being made at Noginsk, and referred to the missile as a 'flying bomb' with a limited range of 400 km.[22] But 'flying bomb' was a term applied to the V-1 cruise missile, not the V-2 ballistic missile. The range was also wrong for the V-2, but right for one version of the V-1, and Soviet copies of the V-2 were not made at Noginsk, thirty miles east of Moscow, but at Podlipki, ten miles north-east of the city. (It is also unlikely, though not quite impossible, that V-2s were being built in the Soviet Union by as early as March 1947.) These considerations suggest that Tokaev simply confused the V-2 with the V-1 in his original narrative.[23] In later versions of this conversation, Tokaev so enlarged and amended his first account as to suggest that Stalin and his advisers had from the outset been almost as interested in long-range *ballistic* missiles as in aerodynamic delivery vehicles.[24] He did this by keeping the reference to 'V-2s' and simply rewriting the rest of Malenkov's remarks. But careful inspection of his original account suggests that in the late 1940s both sides were still preoccupied with aerodynamic options.

A version of part of Tokaev's information about the Soviet rocket programme was published in the *Sunday Express* on 23 January 1949. Although he was interrogated primarily by British intelligence officers, the substance of his story cannot have taken long to reach the Americans.[25] By 1950 reports about the construction of rocket installations at various sites around the Soviet Union were also reaching the United States from other sources.[26]

From 1951 accounts of Soviet missile programmes began to be provided by the returning German experts. However, the Soviet precaution of isolating them from the mainstream Soviet effort now paid dividends, for their limited information failed to disturb a growing American complacency. In August 1952 the results of interrogations of the returnees, conducted jointly by the British Air Ministry and the USAF's Air Technical Intelligence Center, were presented to an important gathering of military, scientific and industrial personnel at the Wright-Patterson Air Force Base in Dayton, Ohio. The Convair representative, speaking with the authority of his company's recent studies for the Air Force, scouted estimates that the Soviet Union would pose a limited ICBM threat to the United States by 1956 and a full-scale one by 1958.[27] In October a report by the Joint Strategic Plans Committee to the Joint Chiefs of Staff noted that 75-mile range American missiles could be expected by 1954, to be followed later by others with ranges of up to 150 miles, and added that in future 'Soviet capabilities will be similarly increased, *though not on a comparable scale as ours.*'[28] A CIA report also remarked that:

> There are numerous reports of guided missile launching sites in many areas of the USSR & Satellites, but to date all such reports are unconfirmed, incomplete, and must be considered unreliable.[29]

Some experts in the United States, however, declined to accept these new and more optimistic assessments of the balance in missile R&D. Walter Dornberger, the former commander of Peenemünde, was working as a consultant at Wright-Patterson and had travelled to Europe to help debrief the German engineers emerging from the Soviet Union. He concluded that the Soviet programme was probably in the lead. That belief was shared at senior levels within the USAF by Major General Donald Putt, at first director of R&D at the newly formed Air Research and Development Command (ARDC) and then its Commander, and his close associate Brigadier General John Sessums. They and a group of like-minded officers believed that the American ICBM programme could and should proceed with more urgency than Air Force Headquarters was then proposing. Citing intelligence about an existing Soviet rocket engine with 264 000 lb thrust, and new Soviet studies for a 550 000 lb engine, Putt warned that 'it is believed quite possible that the Soviets may be engaged in developing an inter-continental rocket missile, and if we delay our Atlas program, we may be running grave risks'.[30]

By the early 1950s, however, this had become, almost inexplicably, a minority view amongst American officials.


## 4 A NEGLECTED HISTORICAL PUZZLE

This survey of a few estimates of the state of Soviet rocket technology that were available to the Truman administration, and that happen to have since been declassified, has raised a serious historical difficulty. With some exceptions the trend appears to have been away from more accurate estimates in the 1940s, which held that the Soviet Union was ahead of the United States, and towards less accurate ones in the early 1950s, which either described the two countries as on roughly level terms, or even put the United States ahead. A possible explanation may simply be that during the Truman presidency rival American estimates of the current state and probable development of Soviet guided missile programmes were never properly reconciled, despite the attempt to remedy such divergences by creating the Central Intelligence Agency in 1947. Early in the Eisenhower administration Trevor Gardner, the Assistant Secretary for Air for R&D, wrote that a survey of available intelligence on Soviet strategic-range guided missiles had yielded 'four separate intelligence estimates, each being substantially different from each other'.[31] A similar overview taken in the late 1940s might well have reached the same conclusion.

It is always hard to explain how knowledge becomes lost or corrupted in the absence of major institutional upheavals. Historians of the American ICBM programme have suggested that intelligence findings were simply adjusted to fit in with changes in the policy on missile development, a process which may be the last thing which should happen to intelligence but which nonetheless occurs.[32] Unfortunately American space historians proper have not been sufficiently aware of the problem to contribute to its solution.

Of course, contemporary historians dealing with any aspect of security affairs are usually hampered by lack of access to relevant secret intelligence documents. With little else to go on, the space historians of the 1960s often referred the reader to an American attitude of contempt for Soviet technological capabilities that was alleged to have prevailed during the Truman period. Thus Robert Perry stated that 'by 1947 it was a precept of American folklore that . . . the Soviet Union was incapable of developing an advanced technology', and von Braun and Ordway wrote that 'Russia was regarded as too backward technologically and too devastated by war to compete with the United States in any field.'[33] The NASA historian Eugene Emme observed that 'the nuclear explosion by the "technically backward" Russians in 1949 was another technological surprise'.[34] There is undoubtedly some truth in such interpretations. Several years after retiring from his post at the head of the American science advice system, Vannevar Bush still maintained that:

> In any field where the Russians can copy a method used elsewhere, they can do so promptly and effectively. . . . But whether they can strike out into new fields and do original things, I rather doubt. I don't think they are that good . . . [35]

Nevertheless, and leaving aside the question of access to secret intelligence for the moment, the first criticism that should be made of the space historians is that their portrayal of post-war American beliefs about Soviet technology has always been far too simplistic. Americans were quite often reminded about Soviet technological prowess, and told that Soviet atomic weapons should be expected by the early 1950s, as for example by contributors to the 1946 symposium *One World Or None*.[36] The immediate and widespread apprehension about the threat which America would shortly face from nuclear explosives in which it held an absolute monopoly at the time, which seized the nation in 1945 and 1946, is evidence that this advice got through to numbers of people. And many ordinary citizens, together with intelligence analysts and senior military officers, were prepared to entertain the possibility that the UFOs which were so widely reported between 1946 and 1950 were some form of Soviet 'secret weapon' or other technology.[37]

The second criticism of the space historians that is supported by the material discussed in this chapter is that, although some of the relevant documents have been declassified in recent years, little has yet been done to correct the erroneous picture of a Truman administration that was entirely deprived of useful intelligence about Soviet rocketry, and swamped with complacent assessments from what one might call the Bush school of thought. This was first developed for entirely partisan purposes by Drew Pearson and Jack Anderson in 1958.[38] But even Walter McDougall has continued the tradition of referring to 'surprising Soviet progress in rocketry', while making no mention of the intelligence documents now available which suggest that another side of the story has yet to be told.[39] And the sentence quoted above from the von Braun and

Ordway *History* has recently been republished, three editions and almost twenty years after its first appearance, as if it were still a defensibly accurate picture of the intelligence situation under Truman and therefore an adequate explanation for his administration's policy, the subject of Chapter 6.[40]

# 6 The Truman Space Policy

## 1 THE POLICY ON BALLISTIC MISSILES

This chapter completes the account of the American rocket programme under the Truman presidency by examining the few occasions on which the undervalued astronautical 'buck' stopped somewhere near that elusive entity, 'the administration'. However, the borderline between generating advice and reaching decisions is often hard to discern in the American political process, especially when the issue is both highly technical and located at the margins of official awareness. Such was the case between 1945 and 1952, both with ballistic missiles and with the artificial Earth satellites which a few informed people began, in the second half of the 1940s, to believe that such rockets would eventually make possible. Because the line separating advice from policy is so hard to draw, it has probably been crossed in one direction in Chapters 3 and 4, and may be crossed again here in the other.

American guided missiles policy under Truman was a minor aspect of air-power policy in general. Air power, in turn, was a controversial subject, focus of intense inter- and intra-service rivalries, which were themselves fuelled by severe funding restrictions, at least until the outbreak of the Korean War in June 1950. In the realm of arms procurement the major issue of the day gradually resolved itself as the development and acquisition of large numbers of nuclear weapons and of aircraft better able to deliver them than the crudely adapted B-29s which had flown to Hiroshima and Nagasaki in 1945.[1]

At the end of 1945 President Truman decided to merge the Departments of War and of the Navy into a 'unified' Department of Defense. The National Security Act was passed in July 1947, after an eighteen-month struggle within the armed services and the Congress. Although it created a separate Air Force, the Act did not assign to it every category of military aviation. The Army, Navy and Marines would each continue to exercise suitable forms of aviation. But the services could not agree upon the necessary definitions and demarcations. The Navy, in particular, disputed the right of the Air Force to monopolize the 'strategic' role of delivering nuclear weapons against remote targets. In December 1948 Secretary of Defense Forrestal summed up the problem tersely in his *First Report*. Noting that there were 'still great areas in which the viewpoints of the services have not come together', he identified the most divisive question as: 'What is to be the use, and who is to be the user of air power?'[2]

Interservice disagreements did not at first affect guided missile programmes as badly as they did the early satellite studies. The Army's Project Hermes, initiated at a late stage of the war with General Electric as the principal contractor, included work on improved versions of the ballistic V-2, using US Army Ordnance personnel at White Sands and the von Braun team of German specialists based at nearby Fort Bliss, Texas. The Navy was mainly interested in short-range guided missiles for air defence. The Air Force had the largest number of missile projects, though all except one were for aerodynamic or 'cruise' missiles.

General Arnold retired in 1946. After that point no one with comparable seniority within the National Military Establishment seemed prepared to stand up for his doctrine that 'the first essential of the air power necessary for our national security is pre-eminence in research', and that a significant portion of presently achievable military strength should therefore be deliberately sacrificed in order to ensure that strength be retained in the distant future also.[3] This was after all a new and difficult idea for a service which had been largely created during the recent war, at a time when investments in potential forms of air power which would only come to fruition a decade or more in the future had been kept to a minimum. As a result there was no effective opposition to the gradualist approach to guided missiles which had been firmly established by the von Kármán reports in 1945.

In particular, the idea that ballistic missiles would somehow be furthered by and arise out of the development of cruise missiles was never re-examined. There was of course some truth in this proposition when it came to propulsion systems, as was shown by the evolution of Rocketdyne engines from the Navaho supersonic ramjet cruise missile to the ballistic Redstone and Atlas systems. Navigation technology could also be inherited, which proved important for early mobile launch platforms like Polaris submarines.[4] But the pre-set guidance systems of ballistic missiles had little commonality with the automatic pilots which controlled long-range cruise missiles. And cruise missile programmes naturally failed to address the problem of re-entry at very high speeds from space into the upper atmosphere, a crucial technology for intercontinental ballistic warheads.

The Army Air Forces R&D chiefs, such as Generals LeMay and Craigie, continued to testify routinely that every part of their missiles development programme was vital. But the reality was a situation in which 'Many people have given lip-service to the magic phrase "Research and Development". Very few of us have really fought for it – and made sacrifices for it.'[5] Such attitudes proved decisive in determining just where the financial squeeze imposed by Truman in 1946 and 1947 would actually take effect. The administration could, in theory, have ruled that nothing be sacrificed from research and development to meet its budgetary requirements. But that was not how the military services were usually governed. Financial restraints were imposed with very little guidance from the top as to how they should be met. Truman's letter of 1 August 1946 to his Secretary for War Robert Patterson, which initiated the process, merely suggested that 'the adjustment should be concentrated in the areas which will release materials and labor resources which are urgently needed to augment production for the civilian economy'.[6] His request was if anything likely to produce harsher results for R&D programmes than if it had not been made. The financial cuts which afflicted Army Air Forces R&D thus represented official policy on guided missiles only in the sense that policy can sometimes be created by omission.

When the Air Materiel Command (AMC) looked for the necessary cuts in guided missile programmes in the summer of 1947 it decided to concentrate resources 'on those missiles which show the greatest promise of early tactical availability'.[7] The only project to develop a truly long-range (5000-mile) ballistic missile, Convair's MX-774 contract, was obviously excluded by that criterion.

The AMC proposals on missiles were then adopted within an overall review of AAF requirements supervised by Brigadier General Thomas Power, the Deputy Assistant Chief of Air Staff for Operations, Commitments and Requirements. Power recommended to the Commanding General of the AAF that long-range surface-to-surface missiles of all types should take fourth place in the list of priorities, after air-launched missiles, short-range (150-mile) surface-to-surface missiles, and surface-to-air missiles.[8] The review was completed in June 1947, and the contract with Convair was cancelled from the start of the next Fiscal Year on 1 July, although rockets completed from existing funds were used for three test launches between then and December 1948. The Air Force re-examined its guided missile priorities in 1949 and 1950, but decided not to change them. At one point it was estimated that the MX-774 would need a further $1.5 million if it was to continue in any form. Arthur Barrows, Under Secretary of the Air Force, refused the request.[9]

The short-sighted, step-by-step approach to the ICBM was further confirmed by the report of the President's Air Policy Commission, chaired by Thomas Finletter, in January 1948. The Commission offered nothing about long-range ballistic missiles beyond such platitudes as 'both [cruise and ballistic] types may prove to be useful, but their characteristics must be greatly improved and their range must be greatly extended'. Even on cruise missiles, the Commission merely repeated the von Kármán line that 'the subsonic missile offers the most practical means of testing and developing the intricate guidance mechanisms for the supersonic types'.[10] The newly-created US Air Force adopted the Finletter Report as basic policy and stuck rigidly to it for at least the rest of the decade.[11]

*          *          *

The former Convair executive Louis Johnson became Secretary of Defense in March 1949. Although he came to feel a sense of urgency about the development of guided missiles in general, and instituted a regular briefing for himself on the progress of the existing cruise-type projects, this did nothing to improve the fortunes of the ballistic MX-774. In September 1949 the RDB produced a Consolidated Technical Estimate which accurately forecast that air-defence missiles would be introduced within five to ten years, and that long-range, surface-to-surface missiles would be ready for testing in 1956, though not for quantity production until some years later.[12] But what they were talking about were still long-range cruise missiles, essentially unmanned aircraft. And indeed the first long-range test flight of a subsonic Snark cruise missile, complete with star-tracking inertial guidance, took place on 14 November 1956.

An Arnold-type policy of 'reducing the force-in-being to enhance the Air Force of the future' was actually proposed in 1949 by a special committee of the USAF Scientific Advisory Board under Louis Ridenour.[13] This was the beginning of a slow and tardy process of change in Air Force policy towards science in general and towards the development of ICBMs in particular. Unfortunately it coincided with the irrational growth of optimism about the relative positions of American and Soviet missile technologies that was described in Chapter 5. The Ridenour Report was endorsed a few months later by a broader military panel under (the former balloonist) General Anderson, commander of the Air University's Air War College, which found that:

> The United States Air Force is now dangerously deficient in its capacity to insure the long-term development and superiority of American power. . . .
>
> . . . Research and Development functions are submerged and diffused in a logistics structure, resulting in the subordination of Research and Development to day-to-day operations. . . .
>
> Current emphasis upon day-to-day operational and materiel problems

has been so great as to radically and adversely affect the long-term development of the Air Force.[14]

The Ridenour and Anderson Reports led to the creation of the Air Research and Development Command (ARDC) in January 1950. However, until April 1951 the ARDC was prevented from functioning independently by resistance from the Air Materiel Command. During this bureaucratic struggle senior ARDC officers were unable to persuade the Air Staff to grant the funds that would be needed to extend a six-month study, which they had assigned to Convair for 1951, into a permanent long-range ballistic missile project.

However, the prevailing lethargy of the Air Force was briefly dispelled in 1949 by anxiety that 'the department of the army is making an all-out bid for control of a disproportionate share of the guided missile program'.[15] The reference was to Army decisions which resulted, the following year, in the transfer of the von Braun rocket engineering team from Fort Bliss to the Redstone Arsenal in Alabama with instructions to develop a 500-mile range ballistic missile. (By that date, it will be recalled from Chapter 5, similar Soviet missiles were already becoming operational.) Early in 1951, with support from the RDB, the Air Force managed to persuade or constrain the Army to restrict the range of such missiles to a maximum of 200 miles. Air Force policy on long-range missiles had come to be that of a 'dog-in-the-manger', neither developing them itself nor tolerating their development by others.

## 2   KAUFMAN T. KELLER

Interservice disputes over missile development projects were possibly exacerbated, rather than curbed, by the intervention of people outside government, which spurred Secretary Louis Johnson and other members of the administration into a stronger interest in guided missile projects. Between May and July 1950 a growing chorus of adverse press comment about waste and poor management in the American missiles programme was ably led by Hanson Baldwin at *The New York Times*.[16] That summer saw the most important decision on guided missiles of Truman's presidency.

On 10 August 1950 Thomas Finletter, who was now Secretary of the Air Force, received the first of two important memoranda from his Under Secretary for R&D, John McCone. The document stressed the probable future vulnerability of the United States to an imminent Soviet capability for air atomic attack, and the consequent need to maintain 'a powerful counter-offensive capacity' whether for war or for deterrence. The direct remedy to vulnerability was to develop 'supersonic ground-to-air missiles for which a crash program should immediately be initiated'. McCone judged the entire American effort in guided missiles to have been poorly organized and seriously underfunded, and recommended a radical re-organization of all missile projects, preferably under a single, centralized body like the wartime Manhattan District. This should be headed by 'the most capable man who can be drafted', who should liaise with a board of directors in the Department of Defense but be directly responsible to the President alone. McCone expanded on the organizational aspects of his proposal in a second memorandum a few days later, and his ideas received prompt support from the Joint Chiefs of Staff and the Secretaries of the Army and Navy.[17]

It should be explained at this point that the resemblance between McCone's ideas in 1950 and those which lay behind President Reagan's Strategic Defense Initiative over thirty years later is by no means accidental. The strategic upheaval brought about by the advent of atomic weapons was perceived at the time as a sudden overturning of the traditional advantage of the defence. This had long been feared as a likely outcome of the development of aerial warfare with conventional explosives, but in fact, with the special exception of Japan, anti-aircraft defences had generally been able to inflict militarily significant losses on attacking air forces during World War 2. In the case of an offensive armed with nuclear weapons, however, such levels of attrition to the bomber force would represent a proportionally far lower strategic 'cost' than that which would be suffered by the defender if even a small percentage of the attack got through. There was consequently a very strong strategic appetite for some kind of 'solution' to the new, and for the hitherto invulnerable United States particularly unwelcome, problem of nuclear defenceless-ness that was being created by the development of long-range bomber aircraft. In these circumstances, the technical possibility, however slight, that air-to-air and ground-to-air missiles might provide the longed-for strategic 'counter-revolution' by enormously increasing the effectiveness of anti-aircraft defence, proved irresistible to the policy-makers of the Truman administration.

On 30 August, after consulting Secretary Johnson, the President invited Kaufman Keller, president of the Chrysler Corporation, to take up the post of an administrative 'czar' with responsibility for all guided missiles and power 'over every building in the capital, the Pentagon included'.[18] But after reviewing the situation Keller concluded that a major reorganization of an effort involving 15 000 people at dozens of military and industrial installations would cause at least a year's delay. He chose instead to act as a part-time consultant, retaining his post at Chrysler, though his actual role was somewhat obscured by the grandiose title of 'Director of Guided Missiles' in the Department of Defense.

On 12 September Secretary of Defense Johnson was suddenly dismissed and General George Marshall, the former Chief of Staff and Secretary of State, was brought out of retirement to replace him. Keller was formally appointed by Marshall on 24 October. Despite his decision to remain a consultant without executive powers, the administration created a public relations impression that something like the wartime atomic weapons programme was being launched for missiles by appointing Major General Kenneth Nichols as his principal assistant. Nichols had been the deputy commander of the Manhattan District, and had then served as a general manager for the Atomic Energy Commission and commanded the Armed Forces Special Weapons Project.

Keller's first report was submitted to the President on 10 July 1951, less than nine months after his appointment. In a covering letter he commented that he was 'amazed at the publicity being given to guided missiles unless it is part of the program of psychological warfare'. One Army and two Navy air-defence missiles were selected for rapid development, although the former (Nike-Ajax) was already at an advanced stage and hardly needed any help from Keller.[19] According to a letter Keller wrote to President Eisenhower in May 1953, while still in his consultancy post, the emphasis on air defence had been mandated both by McCone's original priorities and by Truman's direct instructions.[20]

By February 1951 Keller had endorsed several Air Force missile projects, including the intercontinental but subsonic Snark. Although their ranges, roles, and propulsion systems varied, all were essentially cruise missiles. The only ballistic missile approved by Keller was the Army's surface-to-surface Redstone. In February 1951 he visited the Redstone Arsenal and contributed to the decision to cut the range of Redstone from 500 to 200 miles. This would prevent duplication of the capability about to be supplied by the Air Force's 500-mile Matador cruise missile.[21] But according to Keller's testimony after the sputniks the main reason had been that the expected accuracy of the missile was so low that it was decided to plan for a future 6000 lb thermonuclear warhead and shorten the range.[22]

Keller's first report of July 1951 made no reference at all to ICBMs, and in particular it was silent on Convair's recently completed six-months contract to take another look at the old MX-774 project, now designated MX-1593. The omission was decidedly odd, since he had had extensive dealings with the Air Research and Development Command which had let the contract. At Convair the MX-1593, now starting to be known as 'Atlas', was described obscurely as 'a study and test program leading to the design of a long-range surface-to-surface *tactical* rocket-propelled missile with a primary use as a *strategic* bombardment weapon'.[23] The study funded by the ARDC looked at two concepts in parallel, a pure ballistic and a 'glide rocket' approach, in which a ballistic missile would use wings to facilitate re-entry and possibly also some form of terminal guidance. It was thus in part a continuation of the gradualist approach which had long been favoured at the Air Materiel Command, before the creation of ARDC. It probably also reflected pessimism about the chance of persuading Air Force Headquarters to fund the next stage of the project, that of component development. For at least a year after Convair's report was received in July 1951 neither ARDC nor Convair made any headway in Washington.

After the first sputniks Keller made it clear that he had had strong reservations about the Atlas project, and had in effect suspended work on the propulsion system until further progress was made in laboratory studies aimed at solving the problems of guidance over very long ranges and warhead re-entry.[24] (It was in fact only by breaking with this 'linear' management approach to the problems of ICBM development and substituting the more risky and expensive 'concurrency' approach that Major General Bernard Schriever was able to develop IRBMs and ICBMs at the ARDC's Western Development Division in a few short years after 1955.) Keller's final report went to President Eisenhower in September 1953. It reiterated the emphasis on air-defence missiles and described the Redstone as 'promising', but conveyed his lack of enthusiasm for the Atlas by describing it as 'a highly complex and long term project still in its study stage'. Much like the Air Policy Commission over five years earlier, he warned that 'Over-enthusiasm can also do the program harm.'[25]

$$*\qquad\qquad*\qquad\qquad*$$

The emphasis on strategic air defence and on long-range cruise missiles meant that by the time Truman left office in January 1953 the only intercontinental missiles which the United States had any hope of deploying in the next few years would be pilotless aircraft like the subsonic Snark and Mace or the supersonic Navaho. Long-range ballistic missiles whose propulsion, guidance and re-entry technologies could also be applied to the initial phases of space exploration had yet to be taken seriously in Washington. Thus the world's first operational intercontinental missiles were two squadrons of Snark deployed in Maine in late 1958. (The first operational ICBMs were deployed at Plesetsk in the Soviet Union in 1960.)

Eisenhower's aide Bryce Harlow later recalled that soon after taking office, the President 'looked around and said "Where are the rockets?"'[26]


## 3  POLICY ON SATELLITES

As related in Chapter 4, attempts to coordinate the early post-war satellite studies and proposals through the Research and Development Committee of the Aeronautical Board were not successful. At the end of 1947 the problem was passed to the new Research and Development Board of the Department of Defense.[27]

The RDB took the deadlock over satellite studies seriously enough to address it at their first meeting on 19 December 1947. After they had determined that it should be resolved within the Guided Missiles Committee, the issue was placed in the hands of an *ad hoc* Technical Evaluation Group (TEG) which was chaired by Walter MacNair from Bell Laboratories, and included Francis Clauser, who had edited the Project RAND report in 1946, Richard Porter, a senior General Electric engineer from Project Hermes, Robert Gilruth from the NACA, Guyford Stever from MIT, and Clark Millikan (chairman of the Missiles Committee) from GALCIT. The competence of these experts to pronounce on the technical feasibility of placing a satellite in orbit was beyond question, and on 29 March 1948 they reported that the enterprise was possible with existing or foreseeable technology.

With the exception of Porter, however, TEG members were probably less familiar with the methods and achievements of upper atmosphere research in the ongoing V-2 programme. They determined that:

> . . . neither the Navy nor the USAF has as yet established either a military or a scientific utility commensurate with the presently expected cost of a satellite vehicle. However, the question of utility deserves further study and evaluation.[28]

It was therefore suggested that RAND should concentrate in future on the key issue of the utility of a satellite. But in fact RAND also continued work on system and component design for future reconnaissance satellites, in cooperation with industry.

In September 1948 the RDB's Guided Missiles Committee recommended that Project Hermes should provide 'a continuing analysis of the long-range rocket problem as an expansion of their task on an earth satellite vehicle'.[29] This seems to have been bureacratese for a requirement for the Army to start paying *less* attention to satellite studies and more to calculations about possible long-range missiles.

*       *       *

At this juncture, when technical satellite studies had been virtually suspended and the only project to develop long-range ballistic missiles had also been axed, the military-scientific community and knowledgeable members of the general public were naturally rather surprised to be told that, after all, a coordinated programme of satellite studies was being conducted within the Department of Defense. In December 1948 the *First Report of the Secretary of Defense* contained the following passage:

> The Earth Satellite Vehicle Program, which was being carried out independently by each military service, was assigned to the Committee on Guided Missiles for coordination. To provide an integrated program with resultant elimination of duplication, the committee recommended that current efforts in this field be limited to studies and component designs; well-defined areas of such research have been allocated to each of the three military departments.[30]

Although overall responsibility for the *Report* had lain with Secretary Forrestal, commentators have generally overlooked the fact that this was not really *his* announcement. It was part of a report by Lawrence Hafstad, who served as executive secretary of the RDB (succeeding Lloyd Berkner) from 1947 to 1949, that was included as an appendix. Vannevar Bush, who resigned as RDB chairman on 15 October, may never have seen it, and his successor, Karl Compton, had had little time to check through it. Hafstad was a professor of physics at Johns Hopkins University and had worked on proximity fuses and guided missiles for the Department of War during and after World War 2. He had been familiar with the potential applications of rocketry for launching satellites since the early post-war studies. But it is hard to divine his reasons for publishing a statement so strongly at variance with the fact that service activity on satellites had just been virtually terminated. Perhaps he was a convert to the satellite idea (he joined RAND's Project Feed Back two years later) and simply hoped to generate wider support for it by going public.

Certainly Hafstad would have known whatever there was to know about satellites at the time, so that criticisms of the satellite announcement that were made at a RAND conference on the psychological effects of unconventional weapons a few weeks later were perhaps unfair. Leo Rosten, for example, remarked that 'the worst thing seems to be that the satellite has

been announced, not with bad advice, but without any advice at all'.[31] The
conference report added the pointed observation that:

> Ad hoc announcements or random disclosures may possess lamentable
> significance if viewed from the vantage point of, say, 1955. (The
> conference could not, for example, find a satisfactory answer to the
> question: 'Why was the satellite program announced as it was in
> Secretary Forrestal's First Annual Report?')[32]

Official silence immediately descended upon the subject, and continued
into the following administration. It was broken only in November 1954,
when an obviously reluctant Department of Defense responded to press
enquiries stimulated by the call for satellites to be launched for the
International Geophysical Year (see Chapter 7) with a terse admission
that satellite studies – unspecified – were continuing.

The 'Forrestal announcement' remains one of the most obscure episodes
in the early history of American space policy. The explanation for the
administration's subsequent silence, though not for the original announcement, may simply be that between 1947 and 1949 'the amount of work done
on the satellite program was virtually nil' as RAND's Robert Salter later
stated.[33] On the other hand the Naval Research Laboratory's atmospheric
physicist, Homer Newell, once recalled the existence 'in the late '40s' of:

> . . . the Forrestal Group . . . created in response to a request by Sec-
> retary Forrestal [about] the possibility of launching earth satellites
> and study[ing] why you would do it . . . Harvey Hall was involved
> in that.[34]

It is not clear whether this was meant to apply to the latter part of Forrestal's
time as Secretary of the Navy, during which Harvey Hall's group at BuAer
were indeed conducting such studies, or to his time as Secretary of Defense,
during which Hall's efforts were largely confined to bureaucratic struggles
until June 1948, when the work of his group was officially discontinued.
Hall departed the Navy for RAND a few months later, and it is also just
possible that Newell was referring to satellite utility studies done at RAND
after December 1948, one of the most significant of which was a paper
by Paul Kecskemeti on political and psychological aspects of a satellite
project, completed in October 1950.[35]

*             *             *

In the late 1940s, despite the absence of any actual projects leading towards the development of space technology, the Air Force began to campaign for exclusive rights in space. In January 1948 the Deputy Chief of Staff advised that:

> . . . an Earth Satellite Vehicle is technically, although not economically possible. The passage of time, with accompanying technical progress, will gradually bring the cost of such a missile within feasible bounds. It seems therefore imperative, in order that the USAF maintain its present position in aeronautics and prepare for a future role in astronautics, that a USAF policy regarding Earth Satellite Vehicles be promulgated.[36]

These efforts met with some success in 1949, when the RDB followed its rather scrappy termination of Army and Navy satellite studies in the previous year with a positive ruling that outer space should become the exclusive concern of the Air Force.[37]

Air Force policy was one thing. The policy of the United States should perhaps have been another. But in the late 1940s and early 1950s the latter amounted to little more than a diffuse recognition that, eventually, 'the rocket will be the solution to the satellite problem'.[38] Much therefore depended on the urgency with which large rockets were themselves regarded, for reasons independent of their astronautical applications. We have already seen that at senior levels there was little support for an intensive effort to develop ICBMs. Even at RAND, where satellites had been studied intensively for several years, the early appreciation that a certain amount of haste might be a good idea tended to evaporate with time, perhaps through a process of adaptation to the wider climate of official opinion. In October 1946 James Lipp had used words which anticipated the conversation that Stalin would shortly be having with Tokaev:

> Since the United States is far ahead of any other country in both airplanes and sea power, and since others are abreast of the U.S. in rocket applications, we can expect strong competition in the latter field as being the quickest shortcut for challenging this country's position. No promising avenues of progress in rockets can be neglected by the United States without great danger of falling behind in the world race for armaments.[39]

By January 1949, however, a seminar of experts discussing the psychological aspects of unconventional weapons actually advised against exploiting a satellite project as a demonstration of American technical superiority, on the grounds that:

The appearance of a satellite would only confirm existing Soviet
assumptions concerning *the advanced state of a* [sic] *long-range missile
development in the United States.*[40]

Kecskemeti also assumed without question, in his 1950 discussion paper,
that it would be the United States that launched the satellite, and the Soviet
Union that would have to decide how to respond. He even argued that:

There is a possibility that the launching of a satellite would induce the
Soviets to start a satellite program of their own, possibly for reconnaissance purposes. . . . [However] it is unlikely that the Soviets would
embark upon such a program . . . owing to the ease with which they
can obtain information about United States targets in other ways.[41]

The possibility that the Russians might not need to wait for an American
example, which had been plain enough to Lipp four years earlier, had
suddenly become invisible from Santa Monica.


## 4   ARISTID V. GROSSE

Apart from rumours that the satellite idea was once discussed in Cabinet,
perhaps in the aftermath of the 'Forrestal announcement', the highest levels
of the Truman administration are not thought to have dealt with it directly
until 1952. At some point in that year, perhaps after von Braun launched his
publicity campaign in March, President Truman discussed the matter with
his personal physician and fellow Missourian, Brigadier General Wallace
Graham. Graham apparently persuaded the President to commission a study
of the satellite question from Aristid Grosse, a chemical engineer whose
main experience had been in the petrochemical industry, but who had also
worked as a consultant for the OSRD's first atomic energy studies in the
early 1940s, as well as for other military projects.[42] The substance of
Grosse's eventual report was supplied by 'many hours in conference with
Dr Wernher von Braun'.[43]

Grosse completed his report in August 1953, and sent it to Eisenhower's
Secretary of the Air Force, Harold Talbott.[44] It stated that an unmanned
satellite was now feasible, and 'could be made visible to the naked eye,
under clear atmospheric conditions, at dawn, and at dusk as a bright
fast moving star'. For propaganda purposes, 'with appropriate signaling
or broadcasting devices such a satellite could develop into a highly
effective sky messenger of the free world'. Scientific, reconnaissance or

communications applications would need a concentrated power source; Grosse recommended the use of 'alpha-active radioactive substances', nowadays known as radio-isotopic thermoelectric generators. He warned the successor administration that 'it should not be excluded that the Politbureau might like to take the lead in the development of a satellite'. The Russians might also try to orbit a highly visible object first, and 'the probable reaction of the American people to a Soviet satellite circling about 300 miles above Washington, New York, Chicago and Los Angeles, would have to be considered'. Grosse concluded by recommending:

> . . . that a small but effective committee be set up composed of our top engineers and scientists in the rocket field, with representatives of the Defense and State Departments. This Committee should report to the top levels of our government and should have for its use and evaluation, all data available to our government and industry on this subject. It should report in detail as to what steps should be taken to launch a satellite successfully into outer space and to estimate the cost and time required for such a development. It is felt that if such a committee were in existence and a definite decision taken by our government regarding the construction of a satellite, that it would fire the enthusiasm and imagination of our engineers and scientists and effectively increase our success in the whole field of rockets and guided missiles.[45]

Even if Truman had seen Grosse's report before he left office, he would probably not have adopted a more positive policy towards satellite proposals. Over two years after he had received a copy he still dismissed the American IGY satellite project as 'hooey'.[46] Indeed, perhaps the most significant aspect of this episode is simply that, in spite of the years of careful work on satellites which had been done at the Office of Naval Research and at RAND, by 1952 the President still had to get what appears to have been his first and only impression of the subject from someone acting on behalf of a self-appointed, back-stairs consultant who then took the best part of a year to produce a seven-page typescript containing little but generalities, and handed it in to the administration succeeding the one for which he had supposedly been preparing it.

*          *          *

The obscure and embryonic status to which American space policy was still confined by the end of the Truman administration is also apparent from the transcript of a meeting of an Ad Hoc Committee on Space Flight which was established by the American Rocket Society in 1951. In May 1952

the Space Flight Committee met to review the replies it had received to
a questionnaire about how to improve the prospects for space flight. At
least two of its members felt that things were by then so unpromising in
the United States that the best hope for launching an effective American
space programme might lie in some prior *Soviet* achievement.

Richard Porter, who had been a senior scientist for General Electric on
the Army's Project Hermes and had served on the Missiles Committee's
TEG, suggested that since early satellites would not confer any decisive
military advantage 'the greatest utility with respect to civilization might
be if the Russians were to build it instead of ourselves'.[47]

Kurt Stehling, a rocket engineer with Bell Aircraft, made the same point
with a historical example:

> I suppose we can let our hair down and say that we can and probably
> very definitely may [!] establish an orbital vehicle within five years or
> less. That depends a good deal on other people besides us, that is, across
> the ocean. The Alaskan Highway project, I remember, started in 1940 in
> Canada. People said you couldn't build an Alaskan Highway, that it was
> impossible, and stopped. Suddenly the Japs appeared in the Aleutians
> and we built the Alaskan Highway within a little more than a year or
> a year and a half, over territory than no man in his right mind would
> think a road could be built, a tremendous undertaking, costing over 100
> million dollars.[48]

These remarks were certainly not intended to achieve the degree of
historical irony which in fact they did.

## 5  HISTORICAL TREATMENTS

Harry Truman published his memoirs in 1955 and 1956, before the sputniks, and said nothing about the missile programmes of his presidency.[49]
In 1961 he republished his *New York Times* article about Keller's alleged
achievements.[50] He never tried to defend his administration's record on
the matter of artificial satellites, however, and no one reminded him that
he had dismissed the idea as 'hooey' as late as February 1956. After his
death in December 1972 his friend and physician Wallace Graham replied
to a letter of consolation from Aristid Grosse that:

> President Truman has often spoken about you and your tremendous
> deeds and how with your knowledge and foresight our astronautical

plannings would have been far advanced. It is perhaps I [who] was not able to present this in proper fashion as it was thwarted by the Navy at that time and they gave the President the information that they had plans well in hand.[51]

In view of the understandable personal interest of both Grosse and Graham in the tremendousness of the deeds in question, however, the latter's obituary version of Truman's octogenarian recollections should not receive too much credence.

As for the few space historians who have addressed the Truman record on rocketry and its applications, their usual practice has been to stick to narrative and to eschew explanation. For example, the apparently inconsequential 'Forrestal announcement' about satellite studies in December 1948 has been regularly mentioned, but the only explanation ever offered for the lack of any tangible outcome from it has been the implausible suggestion, first mooted by Cargill Hall in 1963, that official interest in satellite studies ended because of the 'ill will engendered by this report' both in the United States and abroad.[52] The only evidence produced for this theory has been confined to patently biased protests made by *Soviet* apologists.[53] But no one has ever stopped to ask how the alleged tide of domestic public hostility to satellite studies gained such a hold on the Truman administration, by contrast with, for example, the far more vociferous and prestigious, yet still ineffective, protests against Operation Paperclip a year earlier. Still less, how criticisms in the Soviet English-language press were suddenly and uniquely all-powerful with the Defense Department.

Many historians have omitted any reference to Truman's appointment of Kaufman Keller in 1950 to reorganize the missile programme, despite the public discussion of this episode after the first sputniks, which had placed it well and truly on the record.[54] A related tendency has been to portray the ARDC's MX-1593 contract with Convair in January 1951 as if it constituted the decisive re-starting of ICBM development.[55] But at least since Edmund Beard's book came out in 1976, it has been known that this contract was for a six-month paper study only. When the study was completed, the development programme which it recommended was flatly rejected by the Air Force Headquarters, and the residual funding for MX-1593 was set so low that it 'could provide almost no support for hardware development'.[56] There matters were left for the remainder of the Truman administration, by Keller as well as by everyone else. To describe the 1951 study as putting 'the USAF . . . back in the ICBM business', as Walter McDougall has done, is surely to exaggerate.[57]

The first reference in the space literature to Aristid Grosse's investigation of the satellite idea for President Truman appears to have been that published by Shirley Thomas in 1961.[58] The episode has only twice been mentioned by space historians since then, and briefly at that. Neither passage does more than link the text of Grosse's report with the inaccurate information provided by Thomas.[59] As such, they typify the casual approach to the Truman period which has so far been adopted by even the best American space historians.

# Part III International Geophysical Year

# 7 Proposing Satellites for the IGY

## 1 THE AMERICAN ORIGINS OF THE IGY

On 5 April 1950 six American geophysicists and a visiting British colleague at a small social gathering outside Washington conceived the idea that a 'Third International Polar Year' (IPY3) should be organized for 1957–8. The First and Second Polar Years, programmes of internationally cooperative scientific study of various physical aspects of the Earth's polar regions, had been held in 1882–3 and 1932–3, at an interval of fifty years. The idea of holding a Third Year only twenty-five years after the Second did not come entirely out of the blue.

In 1948 Helmut Landsberg alerted the RDB's Committee on Geophysical Sciences to the importance of major global scientific investigations for American security. On 30 March 1949 Lloyd Berkner was appointed by Secretary of State Dean Acheson to direct a military-scientific assistance programme 'in support of the North Atlantic Security Pact' which was on the eve of signature in Washington.[1] In that capacity he supervised the study on *Science and Foreign Relations* mentioned in Chapter 4. The report did not elaborate on the desirability of, or requirements for, cooperative international research projects, which it saw as part of the work of existing international scientific bodies. But it did emphasize that several important sciences needed access to the planet as a whole. Berkner's growing concern with the issues of early warning and strategic defence was expressed in the claim that such work would contribute significantly to the prevention of surprise attacks or threats against the United States, referred to as 'Pearl Harbors'.[2]

In September 1949 *Science* carried an article about the Second Polar Year (IPY2) by John Fleming and Viggo Laursen. Fleming was director of the DTM and president of the Temporary Commission on the Liquidation of IPY2. Laursen was executive officer of the Danish Meteorological Institute, the former director of which, Dan la Cour, had played the leading role in coordinating IPY2. Fleming and Laursen reminded their fellow-scientists that 31 December 1950 was the closing date for the analysis and publication of results obtained from IPY2 experiments, which would be of use to 'scientific investigations for many years to come in which uniform and world-wide simultaneous material is needed'.[3]

The United States had played a larger role in IPY2 than it had in the First Polar Year. It was also the country which published the most scientific reports on the Second Year, probably because scientists elsewhere suffered greater loss of data and of opportunities to analyse them as a result of World War 2. More significantly, IPY2 might never have been carried through at all in the adverse economic conditions of the early 1930s without repeated generous donations from the Rockefeller Foundation.[4]

Marcel Nicolet, the Belgian aeronomist who was to become Secretary General of the IGY, has described the intellectual climate at the DTM and the National Bureau of Standards when he visited them briefly in January 1950, *en route* to the California Institute of Technology at Pasadena. To his recollection, Fleming's influence with American geophysicists went beyond calling for the efficient winding-up of IPY2, to include encouragement to:

> ... consider in what ways a *new* impulse could be given to the development of their discipline. . . . I noticed at once that the ionospheric physicists were particularly anxious to demonstrate their dynamic approach by making many suggestions for the future . . . [5]

Another factor which influenced scientists working on the physics of the upper atmosphere in 1949–51 was the realization that the V-2 atmospheric research programme was coming to an end, and that money for further work with new American sounding rockets, such as Aerobee and Viking, was becoming increasingly hard to obtain. It was a time in which atmospheric physicists, torn between doubts about the immediate military relevance of their research and their financial dependence on the military, were prone to employ what one historian has described as a 'flexible rhetoric' in their attempt to resolve the dilemma.[6] Van Allen has recalled that:

> Around 1950 . . . there was a very rigid rigorous restriction in support for scientific work. Most of us in this high altitude work got the definite impression that opportunities were waning rather rapidly during that period, about 1950, '51.[7]

This impression must have conflicted sharply with the same scientists' estimation, at the time, that:

The fair amount of success enjoyed thus far in the rocket upper atmosphere program is gratifying. Nevertheless, the past years must be regarded as a beginning phase. The ultimate potentialities of the rocket, and perhaps some day the satellite vehicle, are yet to be realized.[8]

\*            \*            \*

From the outset plans for the IGY were dominated by ionospheric physicists, and a major emphasis was placed on the study of physical interactions between the Earth and the Sun. Many of these take place in the upper atmosphere and were difficult to observe until the post-war development of vertically-launched sounding rockets, carrying instruments and telemetric equipment, by means of which short-lived measurements began to be taken above the White Sands Proving Ground and occasionally at other locations. In the circumstances, much of the best work in upper atmosphere physics was being done by Americans and by a few foreign colleagues with access to their results or apparatus. In April 1950 the distinguished British atmospheric physicist, Sydney Chapman, was visiting the DTM. He had recently decided to reduce his commitments in Oxford in order to play a leading role at the newly established Geophysical Institute of the University of Alaska, where he would have better facilities for empirical research. Chapman was particularly interested in some of the first measurements of an atmospheric current system flowing around the planet at 100 km above the geomagnetic equator, the equatorial electrojet, which had been taken by Van Allen and his team at APL with sounding rockets. Van Allen invited him to lecture at APL, and asked his wife to hold a dinner party for Chapman and a few colleagues. During after-dinner conversation the perennial question (for those present) of 'how to obtain simultaneous measurements and observations of the earth and the upper atmosphere from a distance above the earth' was raised.[9] But although that description was based on interviews with some of the participants and may well sound to a modern reader like a reference to space research, it would be unwise to read too much into it. The discussion may not have gone beyond the idea of simultaneous launches of vertical sounding-rockets, such as in fact took place during the IGY. (The first project for multiple sounding-rocket launches over a short period was carried out on 'T-Day', 11–12 December 1950, and some of those present would already have been involved in planning for this.) During the evening someone suggested that a Third Polar Year should be held in 1957–8, twenty-five years after the Second, instead of repeating the fifty-year interval which had separated the first two Years.[10] The group agreed with enthusiasm, and began to discuss how the proposal should be followed through.

In the summer of 1950 two further significant gatherings of upper atmosphere scientists took place in the United States. In May a meeting of about twenty scientists at the Naval Ordnance Test Station at Inyokern, California, including Berkner, Chapman, Nicolet, Van Allen and Joseph Kaplan, the atmospheric physicist who would later take up the post of chairman of the US IGY Committee, supported the idea of laying the IPY3 proposal before the second meeting of the Mixed Commission on the Ionosphere (MCI), a recently formed liaison body intended to coordinate ionospheric research projects of interest to three international scientific unions, the International Union of Geodesy and Geophysics (IUGG), the International Scientific Radio Union (URSI), and the International Astronomical Union (IAU).[11] The MCI was due to meet in Brussels on 4–6 September 1950, in advance of the triennial general assemblies of several international scientific unions in 1951 and 1952, which would be followed by that of their overall coordinating body, the International Council of Scientific Unions (ICSU).[12] Further support for the IPY3 idea was expressed at a conference on ionospheric physics held at Pennsylvania State College in July, which was attended by more than two hundred, mainly American, scientists.[13]

## 2    PREPARING FOR IGY

The published records of the proceedings of ICSU and its affiliates tend to be short on detail, particularly with respect to the business portions of their agendas. The exact terms in which Berkner and Chapman laid the IPY3 proposal before the MCI thus remain unclear. Since Berkner was chairman of the American Committee for Scientific Radio at the time, what Nicolet has called 'the official proposal' for IPY3 may have originated with that body, which held a spring meeting in Washington every year.[14]

The MCI was largely composed of the Anglo-American originators of the proposal and some of their closest colleagues in the Western scientific community, such as the Britons Edward Appleton, Granville Beynon and Harrie Massey, and the Belgians Ernest Herbays and Marcel Nicolet. Chapman was not only esteemed as 'the world's most distinguished geophysicist',[15] but also president of the International Association of Terrestrial Magnetism and Electricity (IATME) and vice-president of the IUGG. It is little wonder that the idea was accepted, together with an initial programme for the Year which was expanded but never radically altered thereafter.[16] The MCI forwarded its resolution to its parent unions, which in turn endorsed it during 1950 and 1951 and notified ICSU to that effect. By January 1951 the Officers Bureau of the ICSU Executive Board had learned of the MCI proposal and given it their interim approval. When the full Executive met in Washington in October 1951 the proposal was officially adopted at the highest level in the international science structure, and an attempt was made to get world-wide preparations for the Year off to a prompt start. This was to be done by appointing a Special Committee of the ICSU Executive for the Year, on the basis of nominations from major international scientific organizations.

In May 1952 the ICSU Bureau completed the formation of the IGY Committee, which became known as CSAGI, for Comité Spécial de l'Année Géophysique Internationale. It consisted initially of Berkner, the Dane Niels Nørlund, the French physicist and vice-president of IATME Jean Coulomb, and the Briton James Wordie, with Colonel Ernest Herbays, then Treasurer of ICSU, acting as secretary/convener, and Berkner as chairman. Beynon, Nicolet and Laursen were named as alternates for Berkner, Nørlund and Coulomb, respectively. Since Nørlund never took up his nomination Nicolet effectively filled that post in CSAGI from the outset.

Formally, it was the Executive Board of ICSU which issued the first invitations to participate in the Year, over the signature of its Secretary General, Archibald Hill. That addressed to the Royal Society was sent on 19 January 1952. It included the information that a proposal (from Chapman) to alter the designation from 'Third Polar' to 'International Geophysical' Year was receiving serious consideration, and indeed referred at one point to 'the Third International Geophysical Year'.[17] At its meeting in February, however, the Council of the Royal Society determined to take no action on the invitation until its delegates had attended the ICSU General Assembly to be held in Amsterdam in October.

The lack of interest of several national scientific bodies in a programme of exclusively polar research was overcome when the ICSU Assembly approved the transformation of 'IPY3' into an IGY, to include programmes of geophysical observation in as many regions of the globe as possible, in particular along three meridians of longitude, 10° East, 140° East and 75° West. Special efforts would also be mounted in the Arctic and Antarctic because of the difficulty and expense of scientific work in those regions under normal conditions, and because of its importance for several of the IGY's principal subjects of research. The Special Committee (CSAGI) was formally elected, with the addition of a third Belgian scientist, Jacques van Mieghem, representing the World Meteorological Organization (WMO). At the end of November 1952, CSAGI's *ad hoc* Officers Bureau sent out a second set of invitations to participate, with requests for detailed proposals for the international research programme, and these 'met a better response'.[18] In the United States the National Academy of Sciences, a non-governmental body, established an IGY Committee in February 1953. Within the next two years four government departments (State, Defense, Commerce and Interior), as well as the government's controlling body for science, the National Science Foundation, were accorded representation on this committee.[19] Nevertheless its relations with government were often difficult, as will be shown in Chapter 11.

At the first CSAGI plenary meeting, held in Brussels in June 1953, twenty-four CSAGI members and observers discussed IGY programme proposals from twelve countries, and called for national programmes to be finalized by the following May. The meeting elected Chapman President of the Year, Berkner Vice-President and Nicolet Secretary General, and from this point onwards the Year was largely run by this triumvirate Officers Bureau.[20] Shortly after the Brussels meeting the ICSU Executive enlarged CSAGI to include seventeen delegates from eight scientific organizations. In 1954 the Committee was restructured by selecting a permanent Reporter for each of thirteen (later fourteen) IGY disciplines or 'subjects' from among its twenty-one members.

*         *         *

Certain features of the preparations for the IGY were especially significant in relation to the development of American astronautics. Important positions within the structures of the Year were held by Americans or by their close associates, some of whom, such as Chapman and Nicolet, had strong professional ties with the United States. Before the Russians came in the French physicists Georges Laclavère and Jean Coulomb were perhaps the most senior figures in the Year with no such connections. Coulomb was appointed a Bureau member in 1956, and two other Reporters were also French, but the French never really became part of the IGY's inner coterie. The tight political control exercised by the IGY leadership was illustrated when the Netherlands IGY Committee proposed to the third CSAGI plenary in 1955 that, in view of the medical and scientific importance of global measurements and studies of the radioactive fall-out from nuclear test explosions, a fourteenth subject of Nuclear Radiation should be added to the programme. At the suggestion of American delegates a very different rationale was created for this proposal, that of dispersing small quantities of radioactive tracers for aeronomic studies. Even so, it was remitted for further preparation. When it was eventually accepted at the 1956 plenary, the position of Reporter was entrusted to Nicolet.

Two key IGY subjects were not distinct scientific disciplines, but coordinated work in all the others. They were that for World Days (which laid down the framework for scientific activity across the entire Year) and that for Rockets and Satellites. The Reporters for both were Americans. In 1955 Britain, Belgium and the United States accounted for eight of the thirteen Reporters. In 1956, when a fourteenth Reporter and the posts of Editor-in-Chief for IGY publications and of an administrative Coordinator for the Year were created, the proportion in key positions rose slightly, to ten out of sixteen.[21]

The coordination of much of the IGY's observational work around a timetable of Regular World Days and World Meteorological Intervals, to which Special World Intervals were to be added at short notice as appropriate, was assigned to the forecasting division of the Central Radio Propagation Laboratory (CRPL) of the National Bureau of Standards, which had opened in 1950 at Fort Belvoir near Washington. Communications between this facility and other centres which it needed to consult were dependent on the services of the US Army Signal Corps, and indeed one historian has recently described the Bureau of Standards itself as having been only 'nominally demilitarized' at the end of World War 2.[22] In the early phases of IGY planning Berkner headed the various committees dealing with World Days. Later his role was taken over by Alan Shapley at the CRPL's headquarters in Boulder, Colorado.


## 3   THE AMERICAN ORIGINS OF THE SATELLITE PROPOSAL

The triennial timetables of the international scientific unions meant that planning for the IGY would go through two main cycles, one in 1951–2 and another in 1954–5. Whether by chance or by design, the notion of trying to use instrument-carrying artificial satellites as part of the IGY programme was not formally mooted until August 1954, early in the second cycle.

Many American scientists participating in the IGY at national or international level, including such prominent figures as Berkner, Kaplan, Newell, Van Allen and Singer, had long been aware of the potential for scientific satellites that was slowly being created by the development of more powerful rocketry in the late 1940s and early 1950s.[23] Towards the end of 1950 Fred Singer, who had attended the Van Allens' dinner party in April, moved to London to take up a post as scientific liaison officer in the European information bureau of the Office of Naval Research. His move may have been a fortuitous and personal decision, but it certainly helped to prepare the ground for the subsequent American proposal that artificial satellites should be attempted during the IGY. During the next four years Singer explained the advantages of small unmanned satellites for investigations of the upper atmosphere to a variety of European audiences, many of whom would later join national IGY committees and delegations to the general consultative meetings of CSAGI.[24]

As was noted in Chapter 4, Berkner later recalled that he had become a convinced advocate of satellites for both strategic and scientific roles by 1952. Nicolet has also stated that by 1953 'the audacious [satellite] proposal was already in the minds of the [CSAGI] bureau'.[25] In August that year all three Bureau members attended the conference on 'Rocket Exploration of the Upper Atmosphere' at Oxford, at which a quarter of the delegates and over three-quarters of the papers were from the United States, including one by Singer on scientific satellites.[26]

In the spring of 1954 both American government officials and key members of the US IGY Committee were reminded of the long-term national interest of the United States in a satellite programme. The RAND Corporation presented the final report of Project Feed Back on 1 March, with a strong recommendation that full-scale development of a reconnaissance satellite should be initiated.[27] A few weeks later President Eisenhower met the Science Advisory Committee of the White House Office of Defense Mobilization (ODM-SAC), which included both Berkner and Alan Waterman, who as director of the National Science Foundation played a leading role in shaping American policy on the IGY. The President directed the scientists' attention 'to the danger of a surprise attack on the United States and . . . the high priority he gave to reducing the probability of military surprise'. This prompted the ODM-SAC to propose, in April, and then to conduct 'a searching review of weapons and intelligence technology'.[28] Although Berkner probably took no direct part in the work of the ODM-SAC Technological Capabilities Panel (TCP), chaired by James Killian, he was certainly well placed to inform himself about it, and in view of his interest in such matters it is virtually certain that he did so. Since the intelligence potential of satellites had been realized in such circles for some time and was a feature of the Panel's subsequent report, completed at the end of the year, it is therefore quite likely that it was already being discussed amongst ODM-SAC members, including Berkner, in the summer of 1954. This argument must remain hypothetical, because the chapter on satellites in the TCP or 'Killian' Report is still classified, almost certainly because it reviewed the political options for the *international presentation* of a satellite programme, rather than because it aired the now familiar technical possibilities of space-based surveillance systems. But according to the National Security Council policy statement

NSC 5520, which was written a few months later, the TCP Report 'recommended . . . an immediate program leading to a very small satellite in orbit around the earth, and that re-examination should be made of the principles of international law with regard to "Freedom of Space"'.[29]

In August and September 1954 Berkner and Singer steered resolutions recommending the use of instrumented satellites for extended observations of atmospheric and solar phenomena through meetings of the MCI and URSI in Brussels and the Hague, and then through the IUGG General Assembly in Rome.[30] The second CSAGI meeting began in Rome immediately after the IUGG proceedings. A late-night caucus of American delegates in Berkner's hotel suite reviewed the satellite idea exhaustively. Singer concentrated on convincing his colleagues that satellites were technically feasible, while Berkner and others emphasized the political desirability of an American 'first' in space.[31] The group concluded that despite the remaining technical obstacles, and despite the fact that, apparently, they had not yet consulted their own government, which would have to fund the project and provide the rocket-powered launch vehicles, they could assure their colleagues from other countries that endorsement of the satellite idea by CSAGI would meet with a favourable response from the United States. Next day, the second CSAGI plenary duly adopted the wording of the IUGG's satellite resolution.

## 4 A CONSPIRACY THEORY OF THE SATELLITE PROPOSAL

In the light of the contemporary scientific background it seems to be the *absence* of any public and official mention of a possible role for satellites in the IGY, before Berkner and Singer launched the idea at the MCI's August 1954 meeting in Brussels, that requires an explanation. Perhaps the overall concept of the Year itself was regarded as a sufficiently challenging and radical proposal for the first available international planning cycle. And even if there was any conscious holding back of ideas about using satellites, that need not be taken to imply that leading American scientists were party to a deliberate conspiracy to finesse their country's satellites into the IGY at the last possible moment before the Russians joined the Year and gained a say in its decisions. In 1950–1 they would have remembered vividly the criticisms of Van Allen's 1948 IUGG paper that were made by journalists and members of the public, and any problems which may have arisen from the 'Forrestal announcement' later the same year. By 1952 they may also have become concerned to avoid being associated with von Braun's sensational publicity for an enormous manned space station.[32]

However, one American scientist who took part in the IGY has stated that from its inception the Year was secretly aimed at securing international acceptance for American satellites. As Chief Scientist at the US Army Signal Corps Research and Development Laboratory at Fort Monmouth, Hans Ziegler was closely involved with technical aspects of the preparation and execution of the American IGY satellite project from the summer of 1955 onwards. In his opinion:

> As preparations for this IGY proceeded . . . the U.S. *gradually intro-duced* information on the evolving potential for . . . satellites. . . . As a result . . . a resolution to recommend the use of satellites . . . was adopted. . . .
> Thus, contrary to common belief that an IGY was routinely coming along for 1957/58 and coincidentally opened a splendid opportunity to introduce satellites as a peaceful international demonstration of scientific applications, it was just the other way around. The U.S. successfully bent the ear of the international groups and cleverly staged an IGY 25 years ahead of schedule, *specifically toward this goal.*[33]

In view of the public-relations disaster which eventually resulted from the American initiative in feeding the satellite idea into the IGY process, this conspiracy theory might claim with more plausibility than most that the protagonists would be unlikely ever to reveal that such an arrangement had existed. Second, Ziegler's claim offers a ready explanation for the continued classification of the final report of Project Feed Back and of the satellite chapter in the TCP Report. It is also a virtual certainty that, if the conversation in the Van Allens' living room really was focused on 'how to obtain simultaneous measurements and observations of the earth and the upper atmosphere from a distance above the earth',[34] then some of those present would have *thought* about using satellites either at the time or shortly afterwards. Satellites may even have been mentioned aloud, although neither Singer nor Van Allen has any such recollection of the occasion today.

On the other hand, while it is of course impossible to prove that such a conspiracy did not develop, either at the further meetings of ionospheric physicists held in the United States in the summer of 1950, or else at a later stage, an allegation as bold as Ziegler's cannot possibly be accepted without hard contemporary evidence, which his memoir simply does not provide.

But to be fair to Ziegler, Berkner's assumption of the role of CSAGI Reporter on Rockets and Satellites at the end of 1954 is rather hard to explain. Unlike the other Reporters Berkner had had no experience in the field which he undertook to coordinate. Homer Newell, who was an expert in the field and in fact deputized for Berkner at several meetings of the CSAGI Rocket and Satellite Working Group, could easily have done the job of Reporter for the subject. This suggests that Berkner was deeply concerned about the political management of this part of the IGY, a hypothesis that will be confirmed in Chapter 8.

## 5  SOVIET PARTICIPATION

In the early 1950s the only international scientific organizations to which Soviet scientific societies were affiliated were the WMO and the IAU. Both took part in the IGY, but neither was much involved in launching it or in drawing up its programme. In general, the Soviet Academy of Sciences did not participate actively in the various ICSU structures until 1955. The position of Soviet allies varied. In the early 1950s several had links with ICSU, but by 1954 only East Germany and Czechoslovakia had joined the IGY, as had Yugoslavia, which was no longer a member of the Soviet bloc. Communist China was absent from most of the scientific unions because it refused to join international bodies unless they expelled Nationalist China.[35]

After announcing at the end of April 1954 that it was considering participation in the IGY, the Soviet Academy of Sciences accepted an invitation to send observers to the IUGG's General Assembly in Rome; formal Soviet adhesion to both IUGG and ICSU followed in 1955. Some of the Soviet scientists stayed on for secret discussions with the CSAGI Bureau and attended sessions of the CSAGI meeting in October 1954, also as observers.[36] At the opening session Chapman announced the news received from the Soviet Embassy that the Soviet Union had officially decided to take part in the Year. At the closing session he invited the senior Russian observer, Vladimir Beloussov, to address the meeting. After confirming that his country had begun to form an IGY committee, Beloussov broached the issue of the representation on CSAGI, as the Year's governing body, of some of the world's larger countries, notably India, China and the Soviet Union. Chapman replied that the IAU and WMO had been soliciting Soviet participation for some time, but pointed out that CSAGI was formed by representatives from the international scientific organizations rather than from countries.

In view of the informal discussions which had already taken place these statements obviously represented an unfinished negotiation. At the

ICSU Executive Board which convened immediately afterwards in Naples Chapman argued that it would in fact be desirable for the Soviet Union to have two delegates in CSAGI, as had the United States in the persons of Berkner and John Simpson, who represented the International Union of Pure and Applied Physics.[37] In 1955 the Soviet IGY Committee was duly registered, and arrangements were made for two Soviet delegates, Beloussov and Nicholas Pushkov, to sit on CSAGI in the nominal capacity of representatives of the IUGG, an organization with which they had hardly had time to familiarize themselves. In 1956 Beloussov and Coulomb were added to the CSAGI Bureau, although neither ever held administrative responsibilities comparable to those of the first three members, and the United States acquired a third CSAGI member when Alan Shapley became Reporter for World Days.

This sequence of events meant that the Soviet Union was never in a position to influence the basic programme of the IGY, which had in essence been written by the US IGY Committee.[38] The Soviet observers at the Rome CSAGI were poorly placed to take part in the discussion of the American delegates' satellite proposal, or to report with confidence on any comparable thinking in a Soviet IGY Committee which had only just begun to be formed and had probably not even met by that point.

The Soviet Union's own planning and policy for a national satellite project appears to have accelerated in response to the CSAGI resolution on satellites. The first step was the establishment, towards the end of 1954, of an Interdepartmental Commission on Interplanetary Communications (ICIC) within the Soviet Academy of Sciences. On 2 August 1955, during the Sixth IAF Congress in Copenhagen, which came immediately after the American announcement that attempts would be made to launch a satellite for the IGY, ICIC member Leonid Sedov spoke to a press conference about a possible Soviet satellite. However, the official version of his remarks contained no indication that the Soviet Union had decided, at that point, to conduct such experiments *under the aegis of the IGY*.[39] An article in the Soviet Academy's journal for September 1955 stated only that the ICIC satellite project would be conducted 'within the framework of UNESCO'.[40] ICSU certainly had close relations with UNESCO, from which it received substantial funding both for the IGY and for its other activities, but the two organizations were quite separate.

It seems probable that Soviet policy about how to present their satellite programme was still undecided by the time of the third CSAGI meeting in Brussels in September 1955. If so, their failure to join in discussions of the satellite idea is fairly understandable, as is the frustration which it induced in their Western colleagues. Whereas five Soviet delegates made up half the attendance at the Working Group on World Days, there was no Soviet contribution to the Working Group on Rockets and Satellites, which contained seven Americans, three Japanese, one Briton (Chapman) and one Australian. Chapman made a pointed remark about 'limited international participation' in the Rockets and Satellites subject, and gave that as his reason for inviting Homer Newell to deliver a plenary lecture on the American satellite project, the text of which was then entered as the only report from the Working Group.[41]

The Rockets and Satellites discipline was not even tabled for discussion at the Eastern European Regional IGY Conference which was held in Moscow at the end of August 1956. As a Soviet-bloc gathering this might have been an appropriate audience for the first announcement of a Soviet IGY satellite project. The fact that the announcement came at the CSAGI meeting in Barcelona a fortnight later suggests that the decision had only just been finalized at that point, as indeed the chairman of the Soviet IGY Committee, Ivan Bardin, explained:

> Since the question of USSR participation in the IGY rocket and satellite observations was decided quite recently, the detailed programme of these observations is not yet elaborated.[42]

## 6   SPACE HISTORIANS ON THE IGY

Few American space writers or historians have ever paid much attention to the political aspects of the IGY which have been discussed in this chapter, such as the leading role of the United States in launching the Year, or the fact that the Soviet Union did not join the Year until after the broad scientific programme had been settled and a call for satellite projects endorsed. Where the IGY has been mentioned, the usual approach has been to describe both the Year in general and the satellite proposal in particular as a worthy and apolitical 'given' launched spontaneously by international scientific bodies. Yet the political balance in the origination of the IGY was repeatedly mentioned in contemporary articles, such as the symposium on the IGY published by the US National Academy of Sciences in 1954.[43] Nor were the relevant facts omitted from the first popular books on the IGY by the Secretary of ICSU, Ronald Fraser, and by the journalist Alexander Marshack.[44]

The American origins of the IGY satellite proposal itself were not widely understood at first, probably because American IGY leaders took great care never to mention that they had initiated it, and instead attributed everything to apparently disembodied international scientific conclaves.[45] Many American space historians have contented themselves with passing on this politically convenient fiction.[46]

That there was no excuse for their having done so is shown by the best early account of the IGY, by the senior *New York Times* science reporter Walter Sullivan, which appeared as a monograph early in 1959 and was expanded to book length two years later. It mentioned both the American origins of the IGY and the late adhesion of the Soviet Union. The book version also began to uncover the background to the satellite proposal by revealing that Singer had moved the resolution to the URSI general assembly in 1954. But Sullivan still omitted all reference to the important part played by Berkner at the MCI, URSI and IUGG meetings.[47]

The first writer on space affairs to consider these aspects of the IGY worth mentioning seems to have been Jay Holmes, who specified that 'the American delegation introduced' the satellite resolution at Rome in October 1954, after a dramatic late-night caucus which he was also the first commentator to describe in print.[48] Amongst NASA historians it was Green and Lomask who gave the fullest account of such matters, in 1971. Until McDougall's book came out in 1985 they were the only professional historians to have noticed and repeated Holmes's description of the American caucus at the Rome CSAGI meeting. They also made it clear that the IGY's overall programme had been written by Americans, that the Soviet Union had not yet joined the IGY when the satellite proposal was accepted, and that 'men interested in the upper atmosphere and space' were prominent in the American national committee.[49]

Unfortunately McDougall's account of the relationship between the IGY and the first satellites does not appear to be based on a re-examination of primary sources. He omits any mention of the activities of Berkner and Singer in the summer of 1954, and twice refers to the CSAGI satellites resolution as if it had been an independent development, before mentioning the American influence which produced it. Nor does he remark such pertinent facts as that the Soviet Academy did not at first declare their satellite project to be within the framework of the IGY.[50]

*        *        *

The national interest of the United States was believed by officials at the time to depend on using a superficially egalitarian programme of international scientific cooperation for the disproportionate benefit, much of it military, of their own country. That arrangement would not succeed, it was argued, if the security benefits to the United States became too obvious. Consistently with this policy the Central Intelligence

Agency declined to take the bureacratic initiative for an American IGY satellite project, though well aware of its significance for their future requirements.[51] The congruence between that conception of the American national interest, on the one hand, and the anodyne accounts of the origins of the IGY and of its endorsement of satellite projects which have usually been rendered by American space historians, on the other, can hardly be viewed as accidental.

# 8  Satellites in the IGY

## 1  PLANNING FOR IGY SATELLITES

As it became clear at a relatively late stage that satellites would be attempted by more than one country during the IGY, the question emerged of what arrangements would be needed for international cooperation. The IGY's 'rules' for satellites were frequently referred to after the sputniks, but their nature and status remain elusive to this day, partly because in some important respects they were never finally agreed between the two main parties, the Soviet Union and the United States, and partly because few Western authors have discussed them objectively. Historical perceptions have been strongly coloured by the extent to which national interests became involved in this part of the IGY, due amongst other things to the security issues associated with rocket launch vehicles when both superpowers were striving to develop the ICBM and to the high stakes that were being played for in the satellite projects, both in national prestige and in career prospects for the scientists and other professional people concerned.

According to a post-sputnik statement by Deputy Secretary of Defense Donald Quarles, American officials somehow interpreted Sedov's remarks in Copenhagen, and in an interview published in the Soviet Union a few weeks later, to mean that an attempt would be made to launch a Soviet satellite *six months before* the opening of the IGY.[1] But in fact Sedov merely referred in Copenhagen to the likelihood of a Soviet attempt 'in the comparatively near future', and in the later interview to 'the possibility of launching an earth satellite in the next couple of years'.[2] How American analysts managed to construe this as predicting a Soviet launch capability by as early as January 1957 remains a mystery. A possible reconstruction is that Sedov was first read, correctly, as referring to a time around August or September 1957; that this was recognized as at least six months earlier than the then estimated date of an American launch capability, which had recently slipped from October 1957 to March 1958; and that incompetent officials then managed to substitute a misleading 'six months before the IGY' for the more accurate 'six months before an American IGY satellite' in their thinking.

Whatever the true explanation, by 1956 American government officials and the nominally independent scientists of the US IGY Committee were extremely anxious to learn whatever they could about Soviet satellite plans. In January Committee staff compiled a dossier of recent material from Soviet sources, but the picture remained frustratingly vague.[3] For officials in the know, concern was doubtless increased by the information about the achievements of the Soviet missile development programme that began to be generated by the CIA's new listening and radar stations in Turkey from 1955. In May 1956 Brigadier General Andrew Goodpaster, Eisenhower's White House staff secretary, noted that intelligence estimates of 'the "time table" of Soviet capabilities' had been altered.[4]

In June 1956, even though the Soviet Committee had not yet announced that it would undertake an IGY satellite project, Hugh Odishaw, the secretary of the US Committee, wrote to the CSAGI Secretary General Marcel Nicolet suggesting that 'most or all . . . of the second day' of the Rockets and Satellites Working Group sessions during the CSAGI meeting to be held at Barcelona in September 'could be allocated to USSR presentation of their effort'.[5] By July Joseph Kaplan, the chairman of the US Committee, was so anxious about 'the problems of international relations in the IGY and especially in the satellite area' that he saw 'the forthcoming Barcelona meeting as a crucial test'.[6] But when Barcelona came around the Soviet scientists were still not ready to respond to American demands. The discussion of satellites was a one-day and one-sided event, and the Russians merely made the first announcement that they would be trying to launch satellites for the IGY, and that details would be given later.

Since there would not be another full CSAGI meeting until well into the IGY itself, the Americans were intensely interested when early in 1957 Soviet scientists began putting out feelers about the possibility of holding a special conference for the Rockets and Satellites discipline. By June the situation was coming to a head. At the beginning of the month the Soviet press reported two statements by the president of the Soviet Academy of Sciences, Alexander Nesmeyanov, referring confidently to the feasibility and desirability of satellites. Nesmeyanov's second statement was particularly striking, since he said that the first satellite would be launched 'soon, literally within the next months' and that 'the apparatus by means of which this extremely bold experiment can be realized *has already been created'*.[7] On 11 June the first test of the first American ICBM, the Atlas, was a failure. At some point in the month the first stereoscopic photographs of a prototype of the first Soviet ICBM were obtained by a U-2 spy-plane which overflew the newly opened second Soviet missile development site at Tyuratam in Kazakhstan.[8] On 22 June the first successful test-flight of a 1000-mile Soviet missile was monitored by the CIA stations in Turkey. Officials in Washington began to fear that a significant Soviet announcement would soon be made, perhaps even to the effect that the world's first satellite had been launched by their superpower rival.

At the beginning of June Berkner took the position that, as the IGY Reporter for Rockets and Satellites, he was not prepared to convene the conference requested by the Russians unless 'they agree to provide a complete and detailed discourse of their technical program on satellites for advance study by other nations'.[9] A meeting of the CSAGI Bureau, of which Beloussov was by now a member, was due in Brussels on 16 June. At that point Berkner's stance produced a response, in the shape of the long-awaited Soviet notification of their IGY satellite project, written in English and dated 10 June.[10] However, Western scientists were intensely disappointed by the vagueness of the Soviet document. Their views were forcibly expressed by Richard Porter, the chairman of the US IGY Committee's Technical Panel on the Earth Satellite Program (TPESP), in notes which summarized the information missing from the Soviet presentation:

*They have not said:– and we have*:

1.    Place of launching, incl. of orbit, alt. of perigee and apogee.
2.    How many? How big? How they will be boosted?
3.    What Govt. agencies or firms are doing the work.
4.    What kind of radio signals will be emitted – frequency, power, polarization, modulation, frequency stability?
5.    Surface coating – reflectivity?
6.    Type of ground radio stations – where located?
7.    Type of photo-telescopes – where located.
8.    Type of telemetry used? Machines to be used for computation.
9.    Any provision for optical back-up?
10.   What data will be obtained?
11.   Types of instruments to be used?
12.   Status of their program? [11]

After the sputniks John Hagen, the director of the American satellite project, told Congress that in the matter of advance presentation of their IGY satellite plans 'the Russians may not have violated the letter of the law in the agreement, but what the Russians did not do was to present their plan as openly and as freely as we did'.[12]

## 2    DEVELOPING THE MANUAL

But what exactly *was* 'the letter of the law' for IGY satellite projects? The arrangements for scientific cooperation in each IGY discipline were encoded in an instruction manual, which usually passed through several drafts as plans were elaborated and additional participants joined in the preparations. The *Manuals* set out the kind of scientific data to be collected and the procedures for transmitting them to and between the three World Data Centres (WDCs), based respectively in the United States, the Soviet Union, and thirdly, in Europe and Japan.[13] They also touched on forms of direct interaction between scientists in different countries.

On 29 November 1956 Sir Archibald Day, the IGY's Coordinator, circulated an outline chapter on Rockets and Satellites in preparation for the *Guide to World Data Centres*. Berkner rewrote this completely by 19 December 1956 to produce the 'rough draft' of a *Manual* which required IGY scientists, among other things, to lodge 'a complete tabulation of the reduced, calibrated, and corrected data derived from the telemetering records of each [satellite] experiment' with a WDC within eight months of launching it.[14] At this stage he envisaged publishing the final *Manual* immediately after a Rockets and Satellites conference to be held at the beginning of August.[15] In the event the *Draft Manual*, described as a 'well-advanced . . . rough draft' in which 'certain deficiencies of information' had yet to 'be supplied for its completion', was further revised in July 1957 and circulated in mid-August.[16] It was then discussed at the actual IGY Conference on Rockets and Satellites, which opened in Washington on 30 September, four days before the launching of Sputnik 1.

On 10 October 1957 Day circulated yet another version of the chapter for the *Guide to World Data Centres*, based on the Washington discussions.[17] But in Moscow in January 1958 the Soviet IGY Committee presented him with an alternative text, containing provisions which the Americans felt to be far too vague. With two satellites already in orbit the Russians claimed that it was still too early for them to decide on the exact form in which orbital data should be presented. Day sent out yet another draft on 28 February 1958 which introduced an obligation to specify the weight of every object placed in orbit and repeated the injunction that the experimental data to be exchanged should consist of precise tabulations and not merely of general interpretations. After objections from several national committees some of the requirements in this version were further relaxed and three paragraphs were dropped, though not the new stipulation about describing every object placed in orbit, which might and in the Soviet case always did include a militarily-significant rocket upper stage. In a fresh version, dated 1 April 1958, their place was taken by several long clarificatory notes, written by Berkner, explaining how the arrangements were supposed to operate. These passages were a decidedly circuitous device. Berkner's foreword to Section XI of the *Guide to World*

*Data Centres*, which was reproduced entire in the *Manual*, clearly shows that he, not Day, was the primary author. But each clarificatory note was presented as an addendum 'entered by the CSAGI Reporter for Rockets and Satellites', namely, Berkner! However, the point was also repeated that the international arrangements for satellites were regarded as subject to yet more 'adjustments' as necessary.[18] Further small changes were agreed at the Moscow CSAGI meeting in July 1958, but in fact throughout the IGY Soviet and American scientists never reached complete agreement about the procedure to be followed in exchanging satellite data.[19]

The question of whether any 'rules' were ever broken is therefore very hard to resolve. The next three sections of this chapter will examine in detail the issues which the continually evolving *Manual*, controlled by an all-American editorial group, attempted to settle, namely, what information a launching state should release about its IGY satellites before or after launch, how data recovered from satellites should be transferred between IGY stations and WDCs, and whether the radio frequency of IGY satellites should be standardized.


## 3   LAUNCH NOTICE AND ORBITAL DATA

At first there were no formal requirements for an IGY committee to publish a detailed description of its satellite programme in advance, though the Americans set an example of open dealing with the accounts of Project Vanguard which they gave at the CSAGI meetings in 1955 and 1956, as well as in various publications.[20] Then the Barcelona CSAGI meeting passed resolutions to the effect that satellite launching authorities should furnish technical information about optical and radio tracking systems, and about experimental telemetry 'as appropriate', and that launch schedules and other planning information should also be released in advance.[21] Whether or not the Russians could realistically have been expected to satisfy every requirement on Porter's list (above), national IGY committees had good reason to hope that at least the first of these resolutions would be adhered to. One national committee could hardly organize observational work in support of another's satellite programme unless it was told something about the type of work and equipment that would be needed.

The US IGY Committee established a network of radio and optical tracking stations in over a dozen countries in support of its own satellite programme, often lending the necessary equipment and sending technicians to help run it.[22] The Soviet Committee took no such measures, even for IGY committees in allied countries such as Czechoslovakia.[23] However, detailed information about radio and optical techniques for satellite observers was published in advance within the Soviet Union, and some of this was also sent to Berkner and to the British IGY Committee.[24] Two senior members of the Soviet rockets and satellites team also attended a symposium on the subject at the Cranfield Institute of Technology in Britain in June 1957, at which they communicated much of the information they were later to put before the Washington Conference.[25]

The *Draft Manual* submitted to the Washington Conference in 1957 called for notice of a satellite launching to be given within one hour on the special rapid communications system which had been set up for the purpose. The conference decided to extend this to two hours, which was little more than the minimum time required to ascertain that a satellite had been established in orbit and to assemble and communicate basic information about it. (According to the BBC Monitoring Service at Caversham, Sputnik 1 was announced by Radio Moscow at 22.22 GMT on 4 October 1957, which would have been about three hours after the launching, and TASS put out a more detailed bulletin at 22.40 GMT.) In April 1958 this requirement was relaxed still further to 'in no case more than 24 hr later'.[26]

It was recognized from the outset that the exchange of information would not extend to details of the rocket launch vehicles, which would be subject to obvious considerations of military security.[27] (This point was later raised by Senator Leverett Saltonstall, who asked why it had been thought necessary to select a 'non-military' launch vehicle for the American satellite, since there would have been no obligation to publish anything about a 'military' one anyway.)[28] In fact the United States published most details of its Vanguard launch vehicle, apart from the composition of the solid fuel in its upper stage. Less was published about the Army's Juno 1 launch vehicle when it was brought into the programme after Sputnik 1, partly because it was based on the Redstone missile and had some technical overlaps with the Jupiter missile, and partly because by then the American commitment to openness within the IGY was distinctly on the wane.

But although the Americans grew belatedly more discreet about their own satellite launch vehicles, they did not become any more willing to concede similar rights of silence to their rivals. As noted above, the version of the *Manual* which Berkner prepared for the Moscow CSAGI meeting required the launching authority to specify the 'size, shape, reflectivity and weight of *all separate objects placed in orbit*'.[29] The scientific motive for such an arrangement was that much research was being done by means of optical tracking of the spent and separated booster of Sputnik 1, or

'1957α1' in astronomical parlance. But when Robert Jastrow claimed at Moscow that such data on *all* orbiting objects were essential 'prerequisites to analysis of observations', he was perhaps overstating the case with the aim of furthering American national interests which had nothing to do with science.[30] The two sides were unable to agree, with American delegates upholding the position stated in their draft and Soviet delegates making the equally extra-scientific suggestion that 'information about the rocket carrier which is necessary to organize its observation should be given by the launching country in the form appropriate to that country'.[31]

They also disagreed vigorously about the amount of information which a launching authority should be obliged to supply about the orbit of its satellite. This was the first item on Porter's list (above), but as his notes make clear it was closely connected with the sensitive question, for the Soviet Union, of the exact location of the secret bases at which the Soviet ICBM was being developed. Few American officials were privy to the information on this subject gained from U-2 flights, and patriotic scientists probably thought that by gaining detailed information on the orbits and launch times of Soviet satellites they would be serving their country as well as science.[32]

The Washington Conference had tried to cater for the Soviet preference for vagueness in such matters with the formula that the launching authority should publish 'complete orbital elements at an instant near the time of launching'.[33] Once again, Berkner resisted this laxity in his April 1958 redraft, which demanded publication of 'the time and co-ordinates of the point at which orbital velocity was first established'.[34] At Moscow, in Berkner's absence, Soviet scientists responded bluntly 'that their computation system is not well suited for the rapid exchange of either approximate or precise orbital elements', and the requirement was relaxed again to publication of precise coordinates of any orbital point.[35]

## 4 INFORMATION FROM SATELLITES

A ground station observing another country's satellite (the important case for international cooperation) might acquire three different kinds of data: visual observations, general characteristics of the radio signals, including tracking calculations, and recordings of the radio signals. Conversely, the launching authority could provide each ground station with information about recommended techniques and the priority in which they should be applied, as well as predictions about the arrival of its satellite within observational range. Since radio observations offered greater access than optical ones to the internal workings of another country's satellite, most of the disputes about data exchange were concerned with the former. The arrangements set out in the *Manual* frequently confused both the types of information in question and the two principal directions in which it could be passed, namely into and out from the two probable launching authorities and their associated Data Centres. This carelessness meant that more was left to inference and subjective judgement than was prudent in the circumstances.

The April 1958 *Draft Manual* recognized that 'raw data, whether film records of optical observations or primary records of radio observations and telemetered signals, are not suitable for exchange'.[36] The context shows that this was supposed to mean only that the launching authority was entitled to 'reduce' its data into meaningful scientific form before depositing them with a World Data Centre, which would then convey them to one or both of the other Centres. The procedure for operating in the other direction was always intended to be very different; as Vanguard director John Hagen explained in 1956 'there is a definite requirement for telemetry reception on a world wide scale'.[37] Thus a few pages before the passage just quoted the *Draft Manual* instructed observing stations to send raw data 'within one day to the primary prediction authority'.[38] The Washington Conference passed a resolution which stressed 'the great value of continuous telemetry, which would facilitate the recording of measured data at all parts of a satellite orbit'. Obviously such data, and not merely those of use for orbital calculations, were supposed to be sent *to* the launching authority by any ground station that had obtained them. But the Washington resolution also advised the Soviet and American committees to provide 'some explanation of the significance of the recorded data in a typical case'.[39] The point was that IGY scientists were not obliged to pub-lish the interpretative codes which alone could transform the telemetered radio signals from satellite experiments into meaningful data. This was a concession to their professional concern for intellectual property rights, and meant that raw data should remain incomprehensible to anyone else who retrieved them from space until the duly interpreted equivalents were sent out to WDCs by those who had placed the experimental package in orbit.

Even before the launching of Sputnik 1 Soviet delegates to the Washing-ton Conference were questioned vigorously about the sort of information they would like to receive from foreign ground stations which observed Soviet satellites by optical or radio means. A list of such questions was drafted on 3 October, a day after Blagonravov had given a fairly detailed account of the future Soviet satellite.[40]

At Moscow a sub-group of the Working Group on Rockets and Satellites was assigned to discuss 'the flow, quantity, content and format of observations and predictions' with a view to 'completing the loop between the observer and the [human] computer'. It was agreed that the launching country should always be 'the primary receiver of observational data', though ground stations would be free to send the material to other computing centres also if they wished.[41] Emphasis was placed on collecting observations from the few ground stations in the Southern Hemisphere, which had access to large portions of satellite orbits that could not be monitored from stations inside the two launching countries.[42] In general, stations should continue to send in their observations until told by the relevant launching authority that they were no longer required.

On the other side of the relationship, the launching authority was expected to send precise orbital predictions to any observing station that requested them. At Moscow Richard Porter pointed out that only one American station was receiving this material from the Soviet IGY Committee whereas 250 wished to do so. The Russians agreed to supply them. However, in the same exchange Porter implied that American predictions were not yet being supplied to Soviet stations either.[43] Lastly, a more relaxed version of the obligation of launching authorities to publish the results of satellite experiments was carefully spelled out. Preliminary reports were to be published 'several weeks after a launching'. 'Full scientific reports' were to appear 'within twelve months after the end of each experiment . . . in literature of general availability.'[44]

The overall impression left by the sometimes confused presentation of the various obligations accepted by IGY participants is that the primary obligation on ground stations was to transmit observations that would 'assist in *tracking* the satellites and improving the *predictions of subsequent orbits*'.[45] Unfortunately their other obligation, to assist with *collecting experimental telemetry*, though indicated, was stated with far less emphasis and clarity. The primary obligations of launching authorities, on the other hand, were to provide orbital predictions to ground stations and WDCs, to send the reduced data from satellite experiments to WDCs, and to publish their full scientific findings in due course.

## 5   THE DISPUTE OVER SATELLITE FREQUENCIES

The collection and transmission of data by IGY participants would have been facilitated by the use of a common radio frequency in satellite telemetering systems, but this was not to be.

The American preference in Project Vanguard was to play safe with the radio tracking of what was expected to be the world's first artificial satellite. That meant using a relatively high frequency which would be both free from ionospheric interference and suitable for the sophisticated phase-comparison technology of the Minitrack tracking system. At least two radio frequencies were considered for the experimental telemetry and Minitrack, namely 108 and 260 megahertz, or megacycles (Mc/s) in the terminology of the day. Finally in July 1956 it was made clear, in answer to a query from the British, that only 108 Mc/s would be used,[46] though in practice each American IGY satellite carried two transmitters, one at 108 Mc/s and the other at 108.03 Mc/s. These were variously powered by mercury batteries and early solar power cells, and the usual arrangement was to pair a relatively high-power short-life transmitter, suitable for Minitrack reception, with another intended for longer use at lower power, which could be reliably acquired only by the Microlock tracking and telemetry system. Microlock had been developed for the Army's Orbiter satellite proposal by the Jet Propulsion Laboratory in 1955–6, and functioned effectively in the Jupiter C warhead re-entry test flights in September 1956, May 1957 and August 1957.[47]

After several descriptions of the developing American satellite project had been circulated in the latter part of 1955 and the first half of 1956 without specifying which frequency would be used, the choice of 108 Mc/s was first generally announced in August 1956, a month before the third CSAGI meeting in Barcelona.[48] But in September Joseph Kaplan gave a paper on Project Vanguard at Barcelona in which he omitted to specify the frequency chosen for American satellites, and merely promised that further details of the radio and optical tracking systems would shortly be made available.[49]

Meanwhile Soviet scientists were deciding to equip the first sputniks with transmitters operating at 20.005 and 40.01 Mc/s. Here too tracking was an important consideration. But the Soviet system was to be based on the less accurate relative frequency or 'Doppler' method, which did not need as high a frequency as Minitrack or Microlock. The shortage of dedicated tracking stations made it desirable to arrange for the participation of large numbers of radio amateurs who would not usually be able to receive high frequencies; also, since the Soviet launch system could place far more weight in orbit than Vanguard, it could carry more powerful transmitters to reach such amateur receivers. Lastly, the Soviet choice of a *pair* of low frequencies was a bold scientific decision to use the differential effects of ionospheric interference on the two wavelengths, greater on 20 Mc/s, less on 40 Mc/s, as a means to study the depth and density of the charged particle layers in the ionosphere by measuring the small delays between acquiring the two signals for observers at known positions relative to the satellite.[50] The principle of such an experiment was understood by American scientists and used on their sounding rockets in the early 1950s. By mid-1955 it was being suggested as a possible satellite experiment, but it was not selected for the IGY.[51]

The Soviet IGY Committee published their choice of frequencies in every issue of the Soviet radio amateurs' magazine *Radio* from June to September 1957. They also sent the information directly to Berkner in a letter dated 16 August 1957, seven weeks before the launching of Sputnik 1, with a copy to the British IGY Committee.[52] However, the Soviet choice of telemetering frequencies was criticized at the Rockets and Satellites Conference when it opened in Washington on 30 September 1957. American scientists charged their Soviet colleagues with having departed from the spirit and indeed the letter of a resolution passed at the Barcelona CSAGI meeting which had recommended:

> . . . that for all IGY satellites the radio systems employed for tracking and telemetering be compatible with those which have been announced at the current CSAGI meeting, in order that the same ground-based receiving equipment can be used throughout.[53]

In closing the Barcelona meeting Sydney Chapman had spoken of agreement having been reached on the 'standardization' of satellite equipment.[54] However, there are several reasons to hesitate before accepting the claim that the Russian choice of frequencies was in breach of an IGY agreement, as all American historians have done. First, CSAGI resolutions were of two kinds, those which were 'resolved' and mandatory, and those which were merely 'recommended' and advisory. The Barcelona resolution was in the second category. Next, as explained above, the record is unclear as to whether the American choice of frequency was actually announced *at* Barcelona. There was certainly no reason why it should not have been, but there seems to be no evidence that it was.

It is also relevant that the solitary Soviet participant in the Rockets and Satellites Working Group at Barcelona, Ivan Bardin, had little understanding of the subject. When the real Soviet experts such as Sedov and Blagonravov began to address the issues raised by committing their satellite programme to the IGY, almost certainly *after* Barcelona, they were unlikely to see themselves as bound by anything to which Bardin had passively consented. Thus at the Moscow meeting in 1958 Alla

Massevich remarked that 'it had taken *one year* to work out some of the problems in this field and many still remain to be resolved'.[55] The implication is that the Russians did not consider themselves to be properly engaged in an equitable discussion of IGY satellite arrangements with the Americans until the summer of 1957, when their first satellite plan was submitted to the CSAGI Bureau, when their proposed frequencies were published and notified to the discipline Reporter, and when preparations were under way for the Washington Conference.

The British Artificial Satellites Subcommittee recorded its disappointment at the Soviet frequency decision at a meeting on 9 September 1957; although ample time in fact remained to build the necessary reception apparatus, its chairman Harrie Massey later explained that because 'there was considerable scepticism about these satellites, no planning about observing them was initiated'.[56] By contrast, the American delay in appreciating which frequencies the Russians had selected, followed by the last-minute distribution of this information only days before the Washington Conference, resulted in considerable indignation, based on the natural but not entirely accurate belief that it had been the Russian choice of frequencies, rather than the American failure to disseminate the notification of that choice as soon as it arrived, which had 'necessitated jury-rigging tracking equipment in haste'.[57] Berkner's inaction on this crucial matter has never been satisfactorily explained. He seems to have given Walter Sullivan the excuse that he was away from his office when the Soviet letter arrived.[58] But with only a few weeks to go before the Washington Conference that absence can hardly have been a long one, and Berkner was not merely avid for any hard information about the Soviet satellite project by that time, but also already well aware, apparently, that 'additional antennas must be added to U.S. type apparatus to see [*sic*] the satellite of the U.S.S.R.'.[59]

Faced with the fact of the Russian choice of different frequencies from the American 108 Mc/s, the Washington Conference recommended both that all countries should establish 'radio-observation stations for frequencies of 20 Mc/s and 40 Mc/s' and, in deference to American sensibilities, 'that all satellites carry means for transmitting radio signals at a frequency high enough to make possible high-precision radio measurements of satellite position'.[60] The British later formed a more positive opinion of the scientific value of the 20–40 Mc/s combination, even going so far as to urge the Americans to use it on their own satellites, and supported a resolution at the Moscow CSAGI meeting which commended the Soviet choice.[61]

## 6   OTHER ASPECTS OF SATELLITE COOPERATION

The coordination of telemetering frequencies was not the only area in which cooperation over IGY satellite projects fell short of the ideal. On the issue of launch notification matters were complicated by an initial Soviet decision that Sputnik 1 was not a full IGY scientific satellite as such, but rather a preliminary test flight intended simply 'to obtain more precise data for the launching of future artificial satellites'.[62] (This Soviet position was reversed about two weeks later.)[63] The Americans acted similarly with respect to their own Vanguard 'test spheres', and became even more chary of making advance announcements after the launching of Sputniks 1 and 2 and the failure of the first Vanguard test-sphere shot on 6 December 1957, thus going back on their previous commitment made at Barcelona in 1956.

On the issue of data transfer William Pickering, chairman of the TPESP Working Group on Tracking and Computation, told the House Committee on Astronautics in April 1958 that eight months since the launch of Sputnik 1 were 'just about up, so that they owe us some data'. Since the Russians had already sent a *Preliminary Report* on the first two sputniks in February, and would obviously be presenting detailed papers at the Moscow CSAGI meeting in July, it is not clear what Pickering thought was amiss. He added that the codes for interpreting American telemetry had been published (supererogatively) whereas Soviet ones had not.[64] At the same hearings James Van Allen was pressed on this point by Representative James Fulton:

*Fulton*: You do not feel that the Russians have broken any agreements by not giving us the results of their Sputnik I and II firings, nor by [not] giving us a code? That is a most unusual statement.

*Van Allen*: I say they have not broken any explicit agreement. I think they have not acted as rapidly as we might have expected a friend to act, but they are within their agreement at the present time. I might say we have not sent yet to the Soviets any information of a substantially revealing character of our findings with our Explorers.[65]

Van Allen went on to explain that the papers in question were to be published the following week and would be sent to Moscow about two weeks after that.

A few days earlier Fred Whipple had told the same hearings that it was an IGY convention to exchange recordings of telemetry, and that 'a good test' of Soviet intentions would be their readiness, or lack of it, to 'transfer to us measurements they have made of the transmissions from our satellites'.[66] However, a summary of the IGY's policy on information exchange in respect of satellites, prepared for the hearings by Hugh Odishaw, omitted any mention of such a requirement.[67]

In general, it may be doubted whether in the immediate aftermath of the first sputniks the comments of American scientists on the conduct of the Soviet satellite programme were always honest or spontaneous. The director of the National Science Foundation, Alan Waterman, was pressing them to draw attention to:

> ... the Soviet's lack of frankness in making available information on their earth satellite, in contrast to our own full cooperation with the IGY program. He [Waterman] has accordingly encouraged our IGY committee to raise this point with the International Committee for whatever action they might wish to take, since such a protest would have maximum force coming from scientists of other nations.[68]

Clearly the 'force' which Waterman was anxious to generate was neither physical nor scientific.

*        *        *

The Moscow CSAGI meeting, at which numerous papers were presented by both sides, followed three months later. One American paper noted that both Soviet and American telemetry had been recorded in Japan and Australia, but not what then became of it.[69] One Russian paper acknowledged the receipt of optical, and two that of radio observations from unnamed foreign ground stations, not necessarily in the West. One of these papers made it clear that coded telemetry from Sputnik 3 was involved.[70]

The minutes of the British IGY Committee show that recordings of telemetry from the first sputniks were made at, amongst other places, the BBC's Monitoring Station at Tatsfield, the Cavendish Laboratory, Cambridge, the Royal Aircraft Establishment, Farnborough, and the Radio Research Station, Slough. In conformity with the IGY arrangements tabulations of the signal characteristics and, later, copies of the actual tapes were sent to Moscow.[71] But the conduct of ground stations built for the American tracking programme may not have been quite so punctilious. When Sputnik 2 was launched on 3 November 1957 the Americans were expecting a Soviet launch and the telemetry was recorded at Australian and American ground stations. Three months earlier Berkner had acknowledged the importance of Southern Hemisphere stations like Woomera (and by implication the US Army Signal Corps stations in South America) for 'obtaining the fullest scientific information from the flight of the Soviet Satellite'.[72] But James Dungey of Imperial College later recalled that the Australian records of Soviet telemetry were deliberately witheld because Soviet scientists 'refused to give the Australians their code' even though nothing in the IGY agreements obliged them to do so.[73] If the Australians had not taken this attitude the Russians might possibly have been able to use the data from Sputnik 2's on-board radiation counter in the higher part of its orbit, above the Southern Hemisphere, to discover the 'Van Allen' radiation belts before the Americans. The relevant Soviet paper to the Moscow meeting made it clear that the satellite data being used were derived solely from passes above Soviet territory.[74] And Van Allen has confirmed that neither American nor Australian recordings of telemetry from the first three sputniks were ever sent to the Soviet IGY Committee, and that Soviet papers on their experiments on board those spacecraft were therefore largely dependent on data obtained from their own ground and ship-board stations.[75]

As for American attempts at decoding Soviet telemetry without the necessary codes, a former Naval Research Laboratory scientist, Charles Johnson, has recalled that:

> . . . the National Security Agency brought down a copy of the Russian telemetry record that they had recorded, and said: 'look, we can't figure out what all this stuff is in here. We know what this is and this is, and they say, well, what are all these little wiggles?' We looked at it, and said: 'well, that's an ion mass spectrometer record. Do you want to see ours?' (laughs) That was the end of that. They went back home happy. They didn't know whether it was some coding of some other kind that they had missed.[76]

The geopolitical circumstances of the Soviet Union, coupled with the Soviet Committee's relative lack of preparation to retrieve satellite data from the extensive portions of any orbit which were inaccessible to their internal ground stations, actually meant that Soviet satellite experiments were more dependent on well defined and accepted procedures for the transmission of recorded signals and of visual observations from foreign ground stations than were the American counterparts. Soviet scientists were therefore keenly interested in the arrangements for data from their satellites to be collected and forwarded by other IGY committees. In January 1958 they told Admiral Day that they were willing to offer orbital predictions on the first two sputniks in return for tracking data from foreign ground stations; in May 1958 Yevgeni Fedorov complained to Day that the February version of the *Draft Manual* was still far too vague about what material observers should be sending *to* a satellite launching authority.[77]

For better or worse, however, the Russians were poorly placed to reciprocate whatever treatment they received from the West over telemetry records. The first two American satellites, Explorer 1 and the 'test sphere' Vanguard 1, broadcast not only tracking beacons but also constant readings from their on-board instruments, a small percentage of which were retrieved by ground stations. The third American satellite, Explorer 3, launched on 26 March 1958, used a relatively secure 'store-dump' telemetry system which recorded data for a whole orbit and could then transmit it in a mere five seconds in response to a signal from the ground. Most (but not all) American ground stations built to recover this material were aligned on a rough north-south axis running through North, Central and South America, on the opposite side of the world from the Soviet Union. Even when the Americans decided to release their codes, as they did for IGY experiments, the Russians were still faced with the problem of placing receiving apparatus on the other side of the planet and switching it on at the right time to record the signals. Thus in the early years of space exploration a combination of geographical inaccessibility, signal encoding and temporal uncertainty was probably enough to protect most early American satellites from Soviet eavesdropping. By contrast, the first four sputniks all broadcast their data throughout their orbits, and the first Soviet satellite with a comparable system for telemetry retrieval was not launched until 19 August 1960.[78]

In early 1958 the Russians declassified the technical details of several of their launching rockets. Later in the year they also released some of their telemetry codes, but this was apparently too late to satisfy their impatient Western critics, who described the action as coming at 'about the time most non-Soviet scientists had given up hope of such Soviet cooperation'.[79]

*     *     *

The degree of non-cooperation in satellite-based IGY research is reflected in a subsequent table of holdings of primary satellite data in the two Western Data Centres for Rockets and Satellites. None of this material was acknowledged to have come from Soviet satellites.[80] When it came to scientific publications the situation at the Data Centres was rather better, with most publications apparently being placed in all three. There was a marked tendency, however, for the authorship of publications to divide along political lines. American and Western bloc scientists did write some papers about the first three Soviet sputniks, but ceased this practice as their own spacecraft arrived and multiplied. Soviet and Eastern bloc scientists seem to have written no papers at all about American satellites.[81]


## 7   THE IGY IN AMERICAN SPACE HISTORY

Few space historians have looked closely at the relationship between the world's first artificial satellites and the IGY framework within which they were conceived. Treatments of the friction between American and Soviet scientists over satellite communications frequencies, for example, have amounted to little more than repetitions of the irritated remarks about alleged Soviet cheating made by American scientists at the time, while the text of Bardin's letter to Berkner announcing the frequencies seven weeks before Sputnik 1 has lain unnoticed in at least one open archive for over thirty years.[82]

Once again, Sullivan provides an exception which puts the work of many later writers to shame. He pointed out that the Barcelona frequency resolution had been only a recommendation, that Bardin 'did not seem familiar with satellite problems and acceded to the various western proposals without argument', and that the Soviet scientists had had good scientific and technical reasons for their choice. He also stressed that the Soviet frequencies had been published in the Soviet Union and communicated to the West well before Sputnik 1, only to be mislaid.[83]

It will be shown in Chapter 11 that because of the domestic political crisis touched off by the sputniks the Eisenhower administration had a strong interest in extracting as much psychological comfort and cold-war convenience as possible by portraying Soviet actions in the first space race as cheating, and their own as irreproachable, and that scientists in the US IGY Committee were willing to play along for their own reasons. But instead of re-evaluating the allegations made by frustrated and sometimes ill-informed American scientists and officials at the time, most historians have merely endorsed the patriotic version of events which those complaints did so much to establish. In thirty years only two books, by Sullivan in 1961 and by Green and Lomask in 1971, have attempted to describe and explain what actually happened.

*          *          *

This chapter, and Chapter 7, have made a start, but admittedly only a start, on assembling the material needed for such a re-evaluation. To recapitulate, the IGY was dominated throughout, and especially in its planning phases, by American scientists and their political allies, and with good reason, as a prominent member of the National Academy of Sciences explained in 1964:

> . . . the proposal for an International Geophysical Year, which was essentially a very interesting and exciting scientific proposal, involved highly significant political considerations and *in many ways became an important tool of U.S. foreign policy.*[84]

The Soviet Union did not join the Year until too late to influence the agreed arrangements. The IGY satellite proposal was also 'made in America'. The late formation of the Soviet IGY Committee together with the sluggish process of policy formation about artificial satellites inside the Soviet Union meant that from October 1954 onwards Soviet scientists could only respond, belatedly, to American initiatives.

The IGY *Manual on Rockets and Satellites*, produced by Americans often frustrated in their search for information about Soviet positions and activities, was often unsatisfactory in its formulation of the code of practice which scientists dealing with artificial satellites, whether as experimenters or as observers, were supposed to follow. The status of the code was also unclear. This left room for unilateral interpretation on both sides. Soviet scientists were much later than their American counterparts in publishing details of their satellite programme, and released less specific information when they did so. However they announced the frequencies they had chosen for Sputnik 1 and communicated them to the relevant IGY official, Berkner, in sufficient time for receiving apparatus to have been prepared in advance. It was not Soviet foul play so much as an administrative blunder (at least) in the United States and a misjudgement in Britain which prevented this from being done. The fact that American complaints about Soviet actions succeeded in distracting attention from the awkward question of which Western scientists and intelligence officials were responsible for such mistakes was surely no coincidence.

The US IGY Committee was laudably open about the contents of its satellite project, if less so about its long-term military significance. But neither the demand for information about Soviet launch sites and launch vehicles, nor the witholding of raw data downloaded by Western ground stations from Soviet satellites, nor the attempts to decode that data covertly, were in any sense legitimate by the standards of conduct which American scientists, politicians, and space writers, affected to hold when criticizing comparable Soviet actions. To date, Western space historians have unanimously overlooked the question of whether the Russians were the only party to fail some of the 'tests' offered by the IGY.

A sensible verdict would be that the lack of cooperation on IGY satellite projects was essentially two-sided. One of the very few Western commentaries to take this position noted that the IGY was only 'ostensibly a fully cooperative enterprise', and that 'considerations arising from rivalry were largely responsible for decisions of both the U.S. and the U.S.S.R. to incorporate a satellite project in their respective national programs for the IGY'.[85] This view will receive ample confirmation in the next three chapters, which seek to locate the American course of action over satellites in the IGY within the contemporary context of Eisenhower's early space policy.

# Part IV  Eisenhower

# 9 Advice for Eisenhower

## 1 ADVICE FROM INSIDE GOVERNMENT

Well before his election as President, Eisenhower was briefed by Oppenheimer and others on the new strategic thinking that was being generated in the summer studies of the early 1950s.[1] To judge by his administration's course of action in 1953 and 1954 he agreed with much of what he heard. In particular, an emphasis on continental defence became 'the principal innovation of the New Look'.[2] There were many reasons for this, not least the first test explosion of the Soviet H-bomb programme in August 1953. But one important factor was undoubtedly the intense scientific effort which began to overcome the 'technical difficulties in the way of an effective continental defense system [which had previously] seemed almost insuperable'.[3]

The Eisenhower administration was anxious to control the costs of some of the new technologies around which its 'New Look' was being constructed. But it did not make the large cutbacks in offensive strategic capabilities which many of the scientists involved in the 1951-2 summer studies had coupled with their recommendations for a new effort in early warning and defence. A further source of strain between the administration and the scientific community were the continued security hearings involving scientists in government work. In October 1953 Senator Joseph McCarthy began an investigation of scientists and engineers working on radar and rocket projects in the Fort Monmouth laboratories of the US Army Signal Corps, as a result of which eight were dismissed and over forty suspended from duty for varying periods.[4] In the spring of 1954, while the Fort Monmouth investigation continued, the Atomic Energy Commission held security hearings on Robert Oppenheimer, who was widely esteemed as a moral and intellectual leader of American physics. Although many distinguished scientific witnesses spoke in Oppenheimer's support, the board of inquiry found on 1 June 1954 that his security clearance should not be renewed.

Senator McCarthy's political star began to wane soon after these events, but they cast a lasting blight on relations between government and civilian scientists. Nevertheless, in the early part of the Eisenhower administration scientists who from personal choice or professional necessity maintained their links with government made a decisive contribution to the development of the new national security policy, symbolized by the start of construction for the Distant Early Warning (DEW) radar system in 1955.

Trevor Gardner, the dynamic, missile-building Assistant Secretary of Defense for Air, encouraged the recently created Science Advisory Committee in the White House Office of Defense Mobilization (ODM-SAC), then the senior science-advice organization in government, to hope that it could become more effective under Eisenhower than it had been in the previous administration.[5] Several members of ODM-SAC including its chairman Lee DuBridge had been active in the recent summer studies, and David Beckler, its long-serving executive secretary, has recalled that the first sign of the Committee's increasing influence came in NSC discussions of the DEW radar line idea during 1953.[6] Thereafter, DuBridge persuaded Arthur Flemming, director of the ODM, to arrange a formal meeting with the President, which took place on 27 March 1954. As related in Chapter 7, Eisenhower raised the problem of surprise attack, which had been the subject of several recent secret studies including two from RAND. As a result of this and further, internal, meetings, ODM-SAC established a Technological Capabilities Panel (TCP), chaired by James Killian, the president of MIT. Eisenhower told Killian that he was 'keenly interested' in the Panel's work.[7] The Panel was widely praised, and Eisenhower built on this foundation by continuing to meet 'his' scientists from time to time.[8] The relationship became even better when the distinguished physicist Isidor Rabi, with whom Eisenhower had become friendly during his two years as president of Columbia University, took over as ODM-SAC chairman.[9]

The Panel reviewed its work at an influential meeting on 24 November 1954 with the President, the Secretaries of State and of Defense, the Director of Central Intelligence and other senior officials, at which it was decided to proceed with the U-2 spy-plane project. A final report on 'Meeting the Threat of Surprise Attack' was presented to the National Security Council in February 1955.[10] The chapter on satellites broadly endorsed an IGY-type approach, which should emphasize the scientific and therefore legally acceptable character of any initial satellite project, but did not examine the technical details of any actual system design.[11] (This was by no means the first advisory treatment of the problem of the international legitimation of overflights by future spacecraft, as it was then perceived against a background of a decade of bitter disputes between the superpowers on the subject of aerial overflights – see section 4 below.)

The advice on satellites in the TCP Report should be seen in the context of the ODM-SAC's general support for the IGY, which was frankly nationalist, even militarist, in tone:

. . . this relatively small U.S. investment could add immeasurably to the effectiveness of civilian and military activities costing hundreds of millions of dollars.

The urgency of carrying out the IGY is apparent when one reflects on the important implications of the IGY program on the well-being and security of our country. . . .

It is clear that a number of highly important advances in engineering have been limited in their practical applications because of insufficient fundamental understanding of related geophysical phenomena, for example, the case of radio communications. . . .

. . . the International Geophysical Year provides a unique opportunity for the U.S. to acquire considerable knowledge with minimum effort by means of international collaboration. Yet our capacity to apply this knowledge to practical affairs is greater than that of any other nation.[12]

This was also the period in which the RAND Corporation produced its recommendation, initially in September 1953 and then in a final report on Project Feed Back on 1 March 1954, that 'the earliest possible completion and use of an efficient satellite reconnaisance vehicle' should be treated as a project of 'vital strategic interest to the United States'.[13] The RAND proposal was accepted within the Air Force by the Air Research and Development Command, which gave it the designation WS-117L in December 1953, and was further endorsed by USAF Headquarters in May and by the Coordinating Committee on Guided Missiles in July 1954. By October the difficult question of possible conflicts between a military satellite project and the various top priority missile projects had been raised at the USAF Scientific Advisory Board.[14]

Perhaps in response to the renewed RAND and Air Force interest a group of people from the Office of Naval Research and the Army Ordnance Corps met on 25 June 1954 to put together a proposal entitled Project Orbiter. This was an intentionally 'quick and dirty' approach, formulated by Wernher von Braun around existing US Army rocketry, in which the science package would have to be very limited.[15] Indeed the original proposal, designated 'Project Slug' at one point by von Braun, was for a 'totally inert' 5 lb body with optical tracking only.[16] Fred Singer's ideas about a 50 lb satellite carrying several experiments were discussed at the early meetings of the group, but were relegated to an indefinite 'second phase' of space exploration.[17] The ONR approved the study phase of the project and appointed Commander George Hoover to direct it in October 1954. After a review in December 1954 it was presented to the Assistant Secretary for

Defense for R&D Donald Quarles and others in January and February 1955, but still offered no more than an inert 5 lb payload.[18]

By March Quarles had received papers on two satellite proposals. One was Project Orbiter, sponsored by the Assistant Secretary of the Navy for Air, James Smith, and the other was an offer from the Naval Research Laboratory to cater for the IGY Committee's proposal, to which they were privy, by adapting the Navy's Viking sounding rocket. Quarles was also aware that the Air Force were seeking support from the Department of Defense for the development phase of a satellite reconnaissance system. At Quarles's prompting Secretary of Defense Charles Wilson ruled on 28 March that no further steps should be taken until Quarles had assessed the rival projects and recommended a clear course of action.[19] Meanwhile the high technical feasibility of Orbiter, canvassed intensively by von Braun, undoubtedly smoothed the path for the US IGY Committee's proposal for an as-yet-unspecified satellite, as it was taken to government during the spring. Nor did Orbiter stand still. That summer, during the competition with the Navy's Vanguard proposal with its sophisticated Minitrack radio-tracking and telemetry system, it was decided to add the radio link later designated 'Microlock' (see Chapter 8) to its specifications.

## 2   ATTITUDES AT THE TOP

Whether or not for the kind of reasons which have just been mentioned, Eisenhower was generally supportive towards the IGY. On 25 June 1954 he released to the press a letter in support of the enterprise addressed to Chester Barnard, chairman of the National Science Board, the governing body of the NSF. On 28 March 1955 his press secretary James Hagerty organized the announcement of the US Antarctic Expedition for the IGY at the White House, four months before staging a similar announcement of the American satellite programme. The President made a further statement in support of the Year on the eve of its commencement.[20]

In March 1955 the president of the National Academy of Sciences, Detlev Bronk, and the director of the National Science Foundation, Alan Waterman, began a two-month process of presenting the US IGY Committee's satellite proposal to government. Eisenhower himself was probably briefed in April.[21] But interestingly his private attitude towards such ideas was almost certainly already positive. A Walt Disney television feature on *Man In Space*, prepared in collaboration with Wernher von Braun, was first shown on 9 March 1955. Eisenhower's response was immediate and enthusiastic. The film's producer, Ward Kimball, recalled that:

Eisenhower borrowed the show to run for the brass in the Pentagon. He called Walt personally to borrow 'Man in Space' and they ran it for a couple of weeks. It was like an educational space primer.[22]

According to Kimball, who naturally heard from Disney just after Eisenhower's call, the President had referred to the prejudices against space exploration of senior military officers as the attitudes of 'stuffed shirts'.[23] This was hardly the language of a President hostile to the idea of space exploration or to science in general, as Eisenhower has sometimes been portrayed by the more intemperate amongst his critics. It seems closer to what might have been expected from a man who was 'reported to have stated his belief (while still a general) in the future of space travel'.[24]

Other senior administration members who supported the satellite idea included Assistant Secretary of Defense for R&D Donald Quarles, Director of Central Intelligence (DCI) Allen Dulles, and Deputy Under Secretary of State Robert Murphy. But in May 1955, just before the National Security Council made its policy determination in favour of an IGY satellite project, neither of the last two departments was prepared 'to initiate further formal action'.[25] Secretary of Defense Charles Wilson was more sceptical as to the need for an American satellite project; as he told reporters enquiring about the possibility that the Soviet Union might be first into space, 'I wouldn't care if they did.'[26] Trevor Gardner was also cautious, and by October 1954 had raised the question of whether a satellite project would interfere with the top priority Air Force missile projects for which he was responsible.[27]

As the satellite project moved into troubled managerial and technical waters during 1956 it also began to draw fire from the Secretary of the Treasury, George Humphrey, and the director of the Bureau of the Budget, Percival Brundage. At this point, Eisenhower is recorded in the unofficial 'Gleason Minutes' of the National Security Council to the effect that:

> . . . he had not been notably enthusiastic about the earth satellite when it had first been considered . . . but that we certainly could not back out of it now. [He] could not imagine the United States having made an announcement that it proposed to launch an earth satellite and then failing to deliver on its commitment.[28]

Perhaps Eisenhower's attitude to science advice in respect of the IGY satellite project was shown most clearly in another exchange at the same NSC meeting. Alan Waterman, the NSF director, had been urging an expansion of the programme from six to twelve attempted launches (see section 5 below). The President countered that if the first two shots were failures it might be wiser to abort even the six-attempt series and '*shift to a different launching vehicle*'. Waterman responded that the scientists were convinced that six attempts were the minimum needed to ensure at least one success. 'The President then said that he surrendered, and certainly would not engage in a fight with all the scientists of the nation.'[29]

## 3   ADVICE FROM OUTSIDE GOVERNMENT

One reason for Eisenhower's positive response to Disney's *Man In Space* programme may have been that he had recently had a personal briefing on the Grosse Report from John Dunning, the Dean of Physics at his erstwhile employer Columbia University and an adviser to the new administration on peaceful applications of nuclear power, though nothing is known in detail about their conversation.[30] Amongst senior members of the National Academy of Sciences advising the administration, such as Lloyd Berkner, attitudes to space policy were quite complex. Three major themes can be identified. There was the usual post-war view that science served United States national interests by enhancing traditional military strength.[31] There was a second more sophisticated thesis that military strength was becoming 'less effective as an instrument of national policy', due to Soviet achievements in the strategic nuclear arms competition, and that the United States should therefore further its national goals by 'seek[ing] other measures to protect and extend the realm of freedom'. This would involve recognizing that 'intellectual attainment . . . may dominate the period immediately ahead as the most powerful single instrument of national policy'.[32]

Berkner also toyed, thirdly, with vague internationalist sentiments about the role of science as a universalizing agent that might reduce the hold of national interests in favour of broad humanitarian goals. Before the IGY had begun or the Soviet Union and its allies had even joined in the preparations, he described it as a process in which 'Tired of war and dissension, men of all nations [*sic*] have turned to "Mother Earth" for a common effort on which all find it easy to agree.'[33]

A contemporary view of American scientists, written by a former Truman science official, saw them as:

> . . . still struggling to reconcile their eighteenth-century devotion to science as a system of objective and dispassionate search for knowledge and as a means for furthering the welfare of mankind in general, with the twentieth-century necessity of using science as a means for strengthening the military power of the United States.[34]

The 'twentieth-century necessity' tended to prevail. By bowing to it the scientists were able to effect a partial reversal, during the 1950s, of Truman's anti-élitist science policy. An inner core of physicists trained and tenured at a small group of institutions, which Killian estimated at no more than 200 people,[35] developed into 'something very close to an *establishment*, in the old and proper sense of that word: a set of institutions supported by tax funds, but largely on faith, and without direct responsibility to political control'.[36]

In this process such liberal internationalism as the scientists may have retained from their pre-war culture was largely subordinated to American national interests, with the aid of the classic imperialist device of 'identif[ying] the good of mankind with the strength of their nation'.[37] Berkner, for example, told the House Committee on Astronautics in April 1958 that:

> . . . when one looks at the general properties of the present Government of the Soviet Union, one supposes that the improvement of exchange of information will inevitably soften the attitude of the Soviet Government which seems to be very hard in some respects at the present time. In the field of science I believe one can say that these attitudes are softer than they are in any other field in which we deal in [*sic*] with the Soviet Union. This, I believe, is an important factor in the relief of tensions.[38]

Although the conspiracy theory that satellites formed a hidden agenda for the IGY almost from its original conception in 1950 must be rejected, the science advice élite in the National Science Foundation and its non-governmental partner the National Academy of Sciences, the parent body to the US IGY Committee, were remarkably responsive to a series of papers on unmanned satellites that were put before them by the Space Flight Committee of the American Rocket Society between 1952 and 1954. The initial ARS satellite proposal preceded Orbiter by almost two years, but that was not the only reason for its preferential treatment by officials of the National Academy and National Science Foundation. It also coincided with the formative period of the American IGY programme, in which it became clear that the IGY would make possible a revival in the fortunes of rocket-based studies of the upper atmosphere, a discipline whose practitioners had long understood and anticipated the potential scientific value of satellites.[39]

Thus an important part of the relationship between the ARS and the scientific élite was the former's commitment, however nominal, to the

principle that the first satellite project should be aimed primarily at
scientific goals, rather than at acquiring data applicable for other purposes,
such as national security. It took the ARS some time to learn that, even
though the former ONR chief scientist Waterman and other leading
scientists close to the administration knew very well that a satellite project
would be of great utility to United States defence programmes, it would get
nowhere with the NSF unless it paid homage to the principle of scientific
primacy. No such thought was present in the Space Flight Committee's
first discussion of the matter in May 1952, at which speculation ranged
freely over political and military aspects of the matter. But by 1953, in a
confidential ARS report submitted to Waterman, scientific utility headed
the list of satellite applications. Finally in 1954, in a public ARS report
which once again urged the NSF to take an initiative, scientific research
was the only application mentioned.[40] But Waterman had apparently had
to intervene directly to stop the Space Flight Committee discussing more
sensitive matters in the manner of a self-appointed alternative National
Security Council. The 1953 report 'was redone after a meeting with
Waterman'. After 'much polishing and rewriting . . . Waterman was able
to use the material in it for his presentation to higher authorities'.[41] Before
that final revision the report contained references to military applications
of satellites; afterwards it did not.

Meanwhile, as related in Chapter 7, the US IGY Committee secured
endorsements for its version of the satellite idea from international scientific bodies, including CSAGI. In January 1955 it 'responded' to its
own initiative with further studies conducted by a special 'Long-Playing
Rockets' (LPR) subcommittee of its Technical Panel on Rocketry. Not
surprisingly, the LPR subcommittee produced a favourable report by early
March. Armed with this Waterman and NAS president Detlev Bronk
proceeded to lobby government, successfully, for a decision in principle
to proceed with an IGY satellite.

*             *             *

In late 1954 and early 1955 the ARS and the NAS could offer an attractive
scientific prospectus for an IGY satellite because in the first case no
particular launch system or payload capacity was specified, and in the
second it was assumed that as much as 50 lb could readily be placed in
orbit.[42] The combination of scientific appeal and technical optimism would
eventually suffice to see off the challenge at the Stewart Committee from
the far more conservatively drawn Project Orbiter. After all, 'utility' had
been a primary consideration in all science advice on satellites since 1948.
The supporters of Vanguard simply took pains to conform to the traditional

view that the national interests of the United States would best be served by a satellite project with the most visible, and actual, utility for science. One of the clearest formulations of the view of the American national interest which triumphed in the space-policy debates of 1955–6 was that given after Sputnik 1 by Eisenhower's first presidential science adviser, James Killian:

> In the long run we can only weaken our science and technology and lower our international prestige by frantically indulging in unnecessary competition and prestige-motivated projects. . . . We shall build greater respect in the long run by ensuring the quality, vigor and integrity of all our science and technology. We shall gain prestige by being better in more areas.[43]

But the Navy's late-entry Vanguard proposal also had the advantage of certain personal connections. Although the IGY Committee's Technical Panel on Rocketry was chaired by Whipple, who had done the optical tracking calculations for Orbiter, its LPR subcommittee was chaired by Milton Rosen, who had directed the Naval Research Laboratory's Viking rocket programme, the ancestor of Vanguard, and who had also led the ARS studies. Another member of the LPR subcommittee, John Townsend, was also from the NRL. Rosen and Townsend started work on the Vanguard proposal even before completing their report from the LPR subcommittee.

Another factor recalled by protagonists from both sides of the Orbiter-Vanguard competition today was a reluctance on the part of American scientists, most of whom had given sterling service to their country during World War 2, to hand the leading role in their country's pioneering space project to a former enemy, von Braun. 'They didn't trust him'; 'they didn't want a German.'[44]

## 4 A NON-MILITARY SATELLITE?

The aspect of the American IGY satellite project which was to attract perhaps the most adverse comment in the national post-mortem on the sputniks was its overtly 'peaceful' character. Before reviewing the advice on which that element of administration policy was based, it is worth pointing out that in the United States in the 1950s a government-sponsored scientific or technological project could be deemed to be directed towards 'peaceful purposes' whether or not the apparatus and personnel used in it were exclusively 'civilian' or 'non-military'. Many civilian scientists were employed by the armed services. And 'peaceful purposes' were by no means expected to exclude the robust pursuit of American political advantage in the Cold War, though outside the policy-making process it was not customary to spell this out.

From a very early stage the secret reports and discussions on satellites at RAND were concerned with possible legal obstacles to American satellite flights passing over the territory of other sovereign nations, some of which would be unlikely to give their permission for such uses of their 'airspace', particularly if they were of a military nature. In his 1946 discussion paper James Lipp proposed the use of an equatorial orbit for the first American satellites, so that they would fly only over territories ruled by governments friendly to the United States, including some of the European colonial powers such as Britain, France and the Netherlands.[45] This was in fact the orbit originally proposed for Project Orbiter, for which tentative plans were drawn up for a launch base at Abemama Atoll in what were then the British-owned Gilbert and Ellice Islands.[46]

By January 1949 the political scientist Harold Lasswell was pointing out in a secret RAND discussion that:

The meaning of legal concepts is how we handle words like 'law'. This handling of the terms may be made to fit any situation. . . .

We can use the satellite as part of a campaign for the development of international law. The emphasis should be put on the 'world community' aspects of the satellite . . . [47]

The meeting recommended that 'the United States, at the highest policy level, be informed of the unique political/psychological opportunities provided by this non-violent instrumentality'.[48] A year later Paul Kecskemeti pointed out that 'the focusing of foreign attention upon the military potentialities of the satellite appears to be undesirable' even if 'few people will believe that a satellite will be made and launched purely for purposes of experimentation and study'. He also favoured an equatorial orbit for the first satellite, and recommended that preliminary publicity should stress 'the scientific aspects of the experiment'.[49]

In the public domain the earliest reference in English to the legal difficulties that might be faced by early satellites was made by John Cooper of McGill University in a lecture given in Mexico City in January 1951.[50] In October of the same year the deputy director of the UN Legal Department, Oscar Schachter, suggested to the First Hayden Planetarium Symposium on Space Flight that:

. . . a principal object of such laws [for outer space] would be to encourage scientific research and investigation. Thus, there would be the idea of free and equal use rather than exclusive possession.[51]

The obvious inference, although Schachter did not draw it explicitly, was that a satellite devoted to strictly scientific purposes would probably not be faced with any serious diplomatic obstacles. This seems to have been the view taken by the ODM-SAC Technological Capabilities Panel in 1954–5. Although the relevant chapter of its Report is still classified, the Panel is known to have recommended a re-examination 'of the principles or practices of international law with regard to "Freedom of Space" from the standpoint of recent advances in weapon technology' and also to have endorsed an IGY-type approach to the first American satellite.[52]

In line with this thinking a group of American officials and IGY scientists, meeting in July 1955 to prepare the ground for the White House announcement of an IGY satellite project, recommended that 'scientific aspects of the venture should be emphasized'. The role of the Defense Department was likened, somewhat naïvely, to its logistic and supposedly apolitical support of polar exploration. Whereas 'the technical means of launching are classified and should not be discussed . . . the bird itself [the satellite] is open to international inspection before launching and in orbit'.[53] In a contemporary interview NSF director Alan Waterman and two scientists from the IGY Committee conceded that the satellite programme 'would be of use militarily insofar as finding out conditions in the upper air for air flight and communications'. Asked whether it 'would . . . help with guided missiles', they agreed that it would.[54]

In general, before Sputnik 1 American IGY scientists made no bones about their project's military connections when addressing international scientific gatherings, and American space writers straightforwardly referred to 'a three-services program', 'the Navy's . . . satellite program', 'the Navy's . . . management of Project Vanguard', and so forth.[55] As Van Allen now points out, although the scientific work of the US IGY Committee was funded through the National Science Foundation 'All of us who worked [on space] in that epoch were sponsored and supported by military agencies, and all the rockets emerged from the military program, even the Vanguard.'[56]

When the NSC decided in May 1955 to proceed with an IGY satellite project, they stipulated that it should not 'materially delay other major Defense programs'.[57] Although the Stewart Committee scientists who took the key decision about which launch-system to recommend were sensitive to this requirement, they were hardly seeking such an improbable goal as a completely non-military one. Von Braun and other officials from Redstone

Arsenal made it clear that, since production of the Redstone missile was being handed over to the Chrysler Corporation and no replacement project had been authorized, far from holding up their work a satellite project would be a welcome task. The majority decision in favour of the Vanguard proposal was based on the technological judgement that Orbiter had a serious reliability problem, with one more stage and many more separate rocket engines than the Vanguard design, each of which could fail, and on the scientific judgement that the Vanguard payload represented the 'maximum scientific utility' that could be attained during the IGY.[58]

It was not, therefore, simply a matter of preferring a 'non-military' Navy system to a 'military' Army one, but of combining the best possible scientific applications with an unavoidably military launch system. As will be shown in Chapter 11, the recommendations of the science advisers settled the line of administration policy in all major respects.


## 5 MOVING THE GOALPOSTS

The US IGY Committee originally proposed a programme of ten satellite-launch attempts. During the final shaping of Project Vanguard in August and September 1955 the number was settled at six 'earnest tries' with the full-sized (21.5 lb) scientific satellite, it being expected that at least one would succeed before the end of the IGY on 31 December 1958. However the scientists discussed a wealth of experimental possibilities at a special conference at Ann Arbor in January 1956, and for the rest of the year they campaigned hard for the number to be raised again from six to twelve, thereby effectively renouncing their original agreement with the administration. Together with the problems of escalating costs and poorly organized funding, this issue came before meetings of the National Security Council on 3 May 1956 and 24 January 1957. On the latter occasion a letter from the ODM-SAC chairman Isidor Rabi was produced in support of an expanded programme, but Eisenhower continued to resist:

> The President replied that he was not averse to the issue [of additional launchings] being brought up any time after we had achieved one successful launching of a satellite. If we do otherwise, we shall simply be gambling on something which it wasn't necessary to gamble on. Our future program ought to be based progressively on what we find out in the course of implementing our present program.[59]

The development programme for the Vanguard satellite launcher was to consist of several Test Vehicle (TV) shots, none of which would carry an actual satellite, followed by firings of the operational Satellite-Launch Vehicle (SLV), which would. But shortly after the NSC meeting just referred to the Naval Research Laboratory team responsible for designing and building the satellite decided to develop a smaller 'test sphere' for use in initial SLV flights, which would give the system a margin of additional power. Also, if such small test satellites could be launched successfully, less system-monitoring equipment would be needed on later full-scale shots, which could therefore carry more experimental apparatus.[60] This idea was presented to the NSC on 10 May 1957 and met with no objection, since it did not appear to extend the number of launch attempts.[61]

On 15 July 1957 the NRL adopted a new version of the test-sphere plan, using some of the TV shots before the SLV launching series proper.[62] Whether or not the NRL scientists had intended all along to use the test spheres to subvert the administration's refusal to extend the number of launch attempts, this is what the scheme now amounted to, though Vanguard's director John Hagen urged his staff 'to consider the test vehicles as test vehicles *only* and that their satellite-launch functions would only be incidental'.[63] The change also meant that a test-sphere launch could be attempted some months before March 1958, by then the earliest expected date for a full-scale SLV shot. This very different second version of the test-sphere plan was never presented to or approved by the administration.

The scientists continued to press for the maximum scientific output from the satellite project, regardless of the need to establish its basic vehicle technology. As Eisenhower for one began to perceive well before Sputnik 1, they were also obstructively indifferent to any other, less intellectual evaluations of what Vanguard meant for American prestige, and this was surely a factor alongside others that contributed to the eventual outcome:

Another problem which disturbed the President was the very costly instrumentation currently being provided for the six satellites. Such costly instrumentation had not been envisaged when NSC 5520 had originally been approved by the President. The President therefore stressed that the element of national prestige, so strongly emphasized in NSC 5520, depended on getting a satellite into its orbit, and not on the instrumentation of the scientific satellite . . . .

The President confessed that he was much annoyed by this tendency to 'gold-plate' the satellite in terms of instrumentation before we had proved the basic feasibility of orbiting any kind of earth satellite.[64]

As these passages show, the administration became increasingly vexed at the spiralling costs of the project. The original estimate, for about ten launching rockets and 'birds' plus five ground stations, had been under $10 million.[65] The first NRL proposal, which covered only costs for satellites and tracking systems, gave the remarkably low figure of $110 000. This probably reflected contemporary assumptions that 'many of the expenses of putting these first man-made moons into the sky will be "painlessly" absorbed by the operating budgets of the military service', or in other words that Vanguard would have unlimited access to what was known to scientists in those days as 'GFM' for 'Government Furnished Material'.[66]

In May 1955, even before the Advisory Group on Special Capabilities (the Stewart Committee) had recommended the Vanguard option, Assistant Secretary of Defense Donald Quarles raised the IGY committee's satellite budget to $15–20 million. The scientists seem to have supposed at first that that limit was the permissible total for NAS expenditures alone, irrespective of those incurred by other agencies.[67] At the NRL, however, a more realistic appreciation of the likely total costs, and also of the fact that it would not be possible to fudge them, began to develop within days of Vanguard's final approval in September. This was because a new and stricter accountancy regime was being introduced at the Department of Defense and throughout the government. The Vanguard estimate rose to $28.8 million in September 1955, $63 million in April 1956, $99.2 million in June 1956, and eventually to $110 million in May 1957, which was a close approximation to the actual final figure.[68]

After the sputniks a Congressional report found that 'the largest single factor' responsible for the cost increases on Vanguard had been 'that realistic estimates of the total program cost were not made initially'.[69] Another part of the financial problem was that, because the administration's original approval for the project had been secured through back-door lobbying by leaders of the scientific élite, no funds had been allocated to the Department of Defense to cover the 'logistics' which it was expected to provide. Yet this included the launch vehicles and the tracking stations, which together accounted for most of the costs. This led to a dispute over financial responsibility between the National Science Foundation and Secretary of Defense Charles Wilson, who had been out-manœuvred by the scientists in their moves to secure the administration's approval for the project. On 1 May 1957 Richard Porter told a subcommittee of the House Appropriations Committee that the launch vehicles were to be funded by the Department of Defense.[70] Nine days later Secretary Wilson was arguing that, because information from the satellite programme would be freely available 'to the whole world', it lacked 'the character of directed research which the Department of Defense could appropriately describe as vital to U.S. national security'. Indeed, he added:

> Congress.had already criticized the Defense Department for allocating money out of its emergency funds to tide over the earth satellite program . . . He complained that he was already having enough trouble in providing money out of his emergency funds for research projects which were truly vital to national defense.[71]

The scientists may to some extent have brought their budgetary problems upon themselves. The original, absurdly low estimate of less than $10 million for a complete satellite project was endorsed by the US IGY Committee in the spring of 1955, after discussions with the Department of Defense, and was almost certainly based on von Braun's Orbiter calculations, which depended on using 'off-the-shelf' rocket technologies and rather basic tracking systems, and on carrying little if any experimental apparatus into orbit. When the Navy's more sophisticated and still-to-be-developed launcher was selected instead, and a range of ambitious experiments began competing for its limited payload space, the costs naturally ceased to have any relation to the original estimate.

Honest miscalculations and changes of plan, with disruptive budgetary consequences, were only to be expected in a novel technological enterprise. As Milton Rosen now recalls, 'We had no good idea what it was going to cost; every estimate I ever made was wrong.'[72] But there is also evidence that the scientists were not entirely averse to misleading the administration about the likely costs of the satellite project. When Army resistance forced a second round of presentations of the rival Orbiter and Vanguard proposals to the Stewart Committee in the second half of August 1955, Rosen reduced the estimated development time for Vanguard from 30 to 18 months simply to match the Army's offer, although privately he still believed the original estimate, which turned out to be correct.[73] Fred Whipple recalls that experience with government contracts since 1945 had taught him that:

> Any honesty in the proposals was disaster. What you had to do was to figure out what it was really going to cost, then you had to put it in as a contingency fund to make the thing sound reasonable. So what I had done on the [Vanguard optical] tracking program was . . . put in the costs that would keep them happy into a tracking program and then put in a very heavy operational budget. . . .
>
> It was not until after the IGY that we began to come in with accurate proposals which would then be accepted. [Before that] there was a

period in there in which an honest proposal would just be turned
down.[74]

But the really culpable aspect of the IGY Committee's original $10
million estimate lay not so much in the fact that the project eventually
cost nearly $120 million (within which several long-term NASA assets
were funded), as in the fact that *nine years earlier* it had already been
estimated by James Lipp at RAND that an experimental, unmanned satellite
would cost between $50 and $150 million, a broad but entirely accurate
estimate.[75]

*        *        *

Another shortcoming in the scientific advice about the satellite project that
was rendered to the Eisenhower administration was its failure to clarify
and emphasize the extent to which satellite-borne experiments would
contribute to the development of fully operational ICBMs. By focusing
on the amount of thrust required to achieve intercontinental distances with
ballistic missiles, together with the fact that it requires slightly more thrust
to achieve a low orbit (with the same payload mass), and by stressing
their undoubted value to pure science, the impression was created that
satellites were a luxury item which could well have been put off until
after the ICBM had been achieved. This was not the case. The geodetic
and atmospheric measurements yielded by the first satellites played a vital
part in ICBM development by clarifying the trajectories that would serve
to join launching and target points, and the environment through which the
warheads would have to pass *en route*. Even the first surveillance satellites,
which had an independent origin in the RAND studies of the early 1950s,
were in some respects beholden to the data generated by the 'scientific'
satellites of the IGY.

Furthermore, although a satellite needed to achieve a slightly greater
velocity than an ICBM, its guidance requirements were between one and
two orders of magnitude *less* demanding,[76] and it did not have to overcome
the problem of surviving re-entry. This meant that useful if primitive satellites could be achieved *before* a fully operational ICBM, the development
of which they would materially assist. The actual development history
followed just those lines. But scientists canvassing for the satellite project
in the early years of the Eisenhower administration failed to make this clear
to senior officials.

Once again, James Lipp had pointed out that development of ICBMs
could benefit from satellites at least as much as *vice versa*, and possibly
more so, as early as 1946. He concluded that:

. . . the satellite will embody many of the problems of control, power plant, staging and construction that will be found in long range rocket missiles. For investigating these features [an equatorial] satellite would be superior to a missile because its repeated passage over the launching point will permit an accurate determination of the results of control and guidance efforts and because *the satellite can furnish continuous data on the upper atmosphere that would be difficult or impossible to obtain with a 'conventional' missile.* If started promptly, the satellite can be of tremendous value in accelerating the long range missile program.[77]

That certain journalists were not unaware of this aspect of the matter was evident in the interview conducted with Waterman and two of the IGY scientists when the American satellite project was announced.[78] But senior policy-makers were left in confusion on this vital point. At the NSC meeting on 3 May 1956:

The President said that it had been his understanding when the Council first adopted NSC 5520, that it had been informed that the successful launching of an earth satellite might be expected to provide information useful to the programs for developing the ICBM and the IRBM. Secretary Wilson, however, pointed out that it was likely to be the other way round, and that the missile programs would be helpful to the earth satellite program rather than the reverse. . . . The President said he judged he was mistaken.[79]

Only in January 1957 did Waterman explain to the NSC 'the beneficial effect on our missiles program of the information we would get from an earth satellite'.[80] Eisenhower had been right and Wilson wrong. But meanwhile the lack of prompt and clear advice on this point had enabled Wilson to muddy the waters at the NSC meeting which assigned the satellite project a priority just below projects on the Master Urgency List of the Department of Defense, by portraying the satellite as something which might provide militarily useful information 'one day' but which meanwhile was only too likely to conflict with ballistic missile programmes.[81]

## 6  HISTORICAL ASSESSMENTS

Immediately after the sputniks most American space writers and other commentators were too busy blaming the policies of the Eisenhower administration for the national humiliation to spare much thought for the pattern of advice which had largely formed those policies. Later historians have generally followed the same line. Eisenhower himself made things easy for his critics by refusing to resort to external or internal recriminations:

> In the circumstances, he could imagine nothing more important than that everybody so involved should stand firmly by the existing earth satellite program which was, after all, adopted by the Council after due deliberation as a reasonable program. In short, we should answer inquiries by stating that we have a plan – a good plan – and that we are going to stick to it.[82]

He even sheltered his advisers by rejecting Detlev Bronk's suggestion that 'if the President were asked if the scientists had been given adequate responsibility and opportunity to develop the satellite, he should say that they had been'.[83]

After the sputniks Eisenhower stressed the part that should henceforth be played by a presidential science adviser in helping to set and then to maintain policy priorities in technological fields. It was tacit acknowledgement that his administration had failed to exert such political control in the past.[84] But a less statesmanlike leader might have reproached his advisers for many things, such as their erratic budgetary submissions, or the disruptive enlargement of the satellite project while it was still an untried technology, or the failure to brief him and other NSC members promptly and clearly about the true relationship between satellites and ICBMs. At a deeper level, Eisenhower was also aware of the part that had been played by the scientists' inability to distinguish between the professional prestige gained through 'intellectual attainment',[85] which was a dominant value within their own culture, and the public, national prestige to be gained by simply launching the world's first satellite, however 'crude' and minimal its instrumentation. Indeed, having publicly committed their government to an IGY satellite project without prior consultation, having selected a launch-system which best suited their élitist and intellectual conception of national goals rather than the view taken by the NSC in May 1955, and having successfuly resisted other policy-making groups which sought to modify the execution of the satellite programme as its likely political outcome began to grow clear, the scientists undoubtedly bore a share of the responsibility for what happened.

*     *     *

Although, as has been said, few writers have ever taken a critical look at the science advice on satellites that was delivered to the Eisenhower administration before the sputniks, there was one direct attack against an individual scientist in 1958. The journalists Drew Pearson and Jack Anderson alleged that Richard Porter's advice had been tainted by corrupt motives.[86]

In 1955 the Naval Research Laboratory selected the General Electric X-405 engine to replace the earlier Reaction Motors engine in the Viking rocket when the latter was upgraded to provide the first stage of the Vanguard SLV. According to Pearson and Anderson, Porter, a General Electric employee, had played a leading part in securing the Stewart Committee's decision in favour of Vanguard. His motive had allegedly been to secure a major rocketry contract for his employers, who were left with much idle plant and unused expertise in the early 1950s after the termination of the Army's Project Hermes, for which they had invested heavily.

The allegation was investigated at hearings of the House Armed Services Committee. Porter testified that he had offered to step down from the Stewart Committee when the Vanguard proposal came before it, but had been dissuaded by the chairman. (Stewart never confirmed this.) Porter also produced a letter from John Hagen, the director of Vanguard, attesting the reliability of the General Electric engine.[87] According to Porter Hagen's testimonial was unsolicited, but a rumour that it had been ghosted by one of the firm's public-relations officers continued to circulate for some years.[88]

*          *          *

One popular writer on space and missile projects, Lloyd Mallan, did in fact publish a more sweeping indictment of American scientists and engineers in 1958, accusing them of timidity and lack of vision with regard to space technology:

> The secret of Sputnik's success – if one must call it a secret – is that Russian scientists were deadly serious about spaceflight when everybody thought they were kidding. The attitude of our own scientists – and the author can state this with firm knowledge, from wide personal experience in talking with top people in the fields of rocketry, aviation medicine and electronics – was simply that flight into outer space is inevitable, but that it would come naturally out of rocket research and development.[89]

Mallan also felt that the lack of an official American organization sponsoring a spaceflight programme was 'not a fault of the U.S. Government' but had resulted from 'the congenital caution of our scientific experts' since 'the President as well as military leaders depend for their actions upon advice of people they consider expert'.[90]

Mallan referred in circumstantial detail to the disinterest felt by the majority of professional rocket engineers in the American Rocket Society, during the late 1940s and early 1950s, towards anything that had to do with spaceflight and space technology. By contrast, writers placing the record of the ARS before the American public after the sputniks never mentioned the possible contribution of its members towards American unreadiness for the Space Age, preferring to follow the easier course of blaming 'the thinking of departments, agencies and commissions in Washington' due to which, allegedly, 'the Soviet Union beat the United States with the first earth satellite . . . because the United States chose to be beaten'.[91]

In 1976 a sociological study of the 'Spaceflight Movement' provided striking confirmation of Mallan's complaints about the conservatism of rocket engineers in general and the ARS in particular.[92] The phenomenon is also evident to anyone glancing through the Society's journal for those years. In November 1953 the minority of space enthusiasts in the ARS were so frustrated at being hived off and largely ignored in the organization's Space Flight Committee that they helped to establish the rival American Astronautical Society.[93]

But space historians and other writers have tended on the whole to accept the American scientific community's simplistic portrayal of itself as timelessly prescient and keen on space exploration. The Stewart Committee's choice of the Vanguard option has been discussed many times, but the details of the advice on potential satellite systems that was given to the Eisenhower administration before October 1957 have never been thoroughly surveyed. After doing so above, it is hard to disagree with the judgement of the doyen of American science advisers, the late James Killian:

I have always felt that the National Security Council and Eisenhower were not given sound enough recommendations by the civilian scientists who argued for the Vanguard project and other approaches to satellites.[94]

However, this view of the matter has yet to convince the space historians.

# 10 Intelligence for Eisenhower

Historians have frequently stated or implied that before Sputnik 1 the Eisenhower administration either failed to provide itself with adequate intelligence of Soviet intentions and capabilities in respect of artificial satellites, or else ignored such intelligence from motives of complacency or parsimony. There are two problems with assessing this, hitherto the majority historical verdict in the matter. The first is that few authors have taken care to distinguish their own hindsighted policy recommendations from their account of the facts. There has been a tendency to assume that, since the administration 'should' have done X in response to intelligence of Soviet satellite capabilities, whereas in fact it did Y, it therefore either had no such intelligence or else was ignoring it. Such reasoning irrationally disregards the possibility that the administration both had the intelligence and took it seriously, but determined on balance to pursue a course of action other than that favoured by the commentator. The second difficulty is that much relevant documentary evidence is still classified. However some is already available. It provides a basis for at least an initial treatment of this aspect of Eisenhower's early, pre-sputnik space policy.

## 1 THE MISSILE GAP PERCEIVED

The first sign that the Eisenhower administration was aware of the United States' comparative disadvantage in respect of rocket technologies was simply the new, more aggressive policy on rocket and missile development which the administration adopted shortly after taking office, despite considerable service opposition and despite its earlier campaign commitment to make major cuts in all types of defence expenditure.

In January 1953 the 200-mile Redstone was the only American ballistic missile with a range beyond the immediate infantry battlefield that was under active development. The Atlas ICBM project, which had been briefly revived at study level during 1951, was once again under leisurely review and virtually at a standstill. Harold Talbott, Eisenhower's first Secretary of the Air Force, appointed Trevor Gardner as his Assistant Secretary for Air for R&D, and in April 1953 Gardner requested an urgent Air Force review of long-range missile projects. It was already starting to become clear from scientific advice that the development of thermonuclear bombs would lead to relatively lighter and more powerful nuclear warheads, and that the high throw-weight and accuracy requirements with which the Atlas project had been burdened could therefore be significantly relaxed. But Gardner became dissatisfied with the cautious approach of the Air Research and Development Command, and still more so with the persistent delaying tactics of the Air Force staff. In November he went outside the existing structures of the USAF Scientific Advisory Board to establish the Strategic Missiles Evaluation Committee (SMEC), a panel of distinguished scientists chaired by John von Neumann. Even before he received their recommendation for an accelerated ICBM programme to exploit the new possibilities in thermonuclear technology, Gardner compounded his break with accepted Air Force procedures by preparing to establish an independent missile-developing agency, albeit nominally under Air Force command. This became the Western Development Division of the ARDC, operating in tandem with a newly established firm of civilian consultants, the Ramo-Wooldridge Corporation.[1]

These developments were accompanied and stimulated by a gradual transformation in the intelligence picture of Soviet missile programmes. It was shown in Chapter 5 that although fairly accurate and therefore pessimistic estimates of the relative rocket and missile capabilities of the Soviet Union and the United States were generated in the immediate post-war years, they tended to be replaced with more optimistic and less accurate ones in the early 1950s. In 1952 reports of new Soviet launch-sites were still being discounted, and the assessment that American missile capabilities were set to increase more rapidly than their Soviet counterparts was still receiving official credence. Under the new administration that kind of estimate began to disappear. By the end of 1953 the SMEC had received 'the lump impression' (from divergent estimates) that 'the Soviets are significantly ahead of us in the strategic missile field'.[2]

In February 1954 the Central Intelligence Agency produced an evaluation of the likely Soviet threat to the continental United States that would be operational by the end of 1957. The conclusion was that so far there were no Soviet RBMs or CBMs missiles in series production and that, for the period in question, no part of the threat would be based on long-range missiles.[3] It was an accurate forecast of intercontinental force *deployments*, but was not inconsistent with the SMEC assessment of the relative positions of the two countries' missile *development* programmes, noted above.

The vague and unreliable revelations about Soviet rocketry in Tokaev's first book and the newspaper article which preceded it, in 1950–1, were gradually superseded by better-informed publications in the next two years.[4] In 1953 and 1954 George Sutton, a senior rocket engineer in the Aerophysics Department of North American Aviation who was described a few years later as 'very much involved in the intelligence activities of his company',[5] published what were probably then the most comprehensive open accounts of post-war Soviet rocketry in the West.[6] After the first sputniks CIA officials testified (below) that the Russians first tested a 220 000 lb rocket engine in 1951. Sutton's first article was based on a lecture given to the American Rocket Society on 4 December 1952, and overall his information closely resembled the data on Soviet rocket programmes that were being cited by Major General Donald Putt at the ARDC in March 1952 (Chapter 5) and therefore, probably, the kind of intelligence about the Soviet missile programme that impressed the members of the SMEC. Sutton referred to a Soviet 'Model 103' development project for 'a giant-size, liquid-propellant rocket engine [with] the tremendously high thrust of 264 000 lb at sea level'.[7] In fact, the RD-103 engine had a thrust of only 78 400 lb, and was used in the Pobeda MRBM which entered service in the early 1950s, well before any Western equivalent. The description better fits the RD-107 engine which formed the central booster for the first Soviet ICBM, the SS-6, as well as for its satellite-launching variant, or the similar RD-108 engine which was used in the four strap-on boosters for the same applications.

The ODM-SAC's TCP Report, which was prepared in the summer and autumn of 1954, did not foresee any Soviet lead in the development of strategic-range delivery capabilities. It distinguished between an early phase of the strategic balance, during which the United States would enjoy 'a very great offensive advantage' based on nuclear-armed bomber aircraft until at least 1958, and a later period of stalemate based on mutual possession of unstoppable ICBMs. The probable consequences of one or the other side's being the first to deploy ICBMs were made clear, but no attempt was made to determine *which* side was actually ahead in the race to do so. In addition to recommending that a major effort be made to develop the ICBM, the Panel also favoured the prompt development of Intermediate-Range Ballistic Missiles (IRBMs), an idea which at least implied that American ICBMs might not be ready in time to counter the threat from their Soviet equivalents. IRBMs based in allied countries closer to the Soviet Union would meet that requirement, as well as countering the developing Soviet missile threat to Western Europe.[8]

The TCP Report added a somewhat retrograde suggestion that 'we should push all promising technological development so that we may stay in [pre-ICBM] Periods II and III-A as long as possible'.[9] This was an optimistic reference to ideas about defensive anti-missile weapons, or AICBMs as they were then called. It was also an unwitting acknowledgement of the point which had been made by James Lipp in 1946, that the race in missile technologies was one in which the United States was at its most vulnerable, since 'other countries' had started out on level terms after World War 2, whereas in aviation and naval power the United States had been generally superior to her competitors at that point.[10]

Early in 1954 the Department of Defense took an internal decision to proceed with the development of the ICBM as a high priority. In September 1955 the SMEC and TCP recommendations were endorsed by the National Security Council and by the President, who approved NSC Action 1433a giving the ICBM programme the highest priority in the country, 'in view of known Soviet progress in this field'.[11] In December Eisenhower directed that IRBM programmes should also have top priority alongside the ICBM, on the grounds that:

> . . . the political and psychological impact upon the world of the early development of an effective ballistic missile in the 1000–1700 mile [IRBM] range would be so great that early development of such a missile would be of critical importance to the national security interests of the United States.[12]

That month also saw publication of the first conjectural table of Soviet missile types by Alfred Zaehringer,[13] and in March 1956 *Aviation Week* took up the point that 'Soviet advances force U.S. missile pace'.[14]

                    *               *               *

The decisive factor in resolving the intelligence picture on Soviet missile projects was the establishment of radars and other technical intelligence-gathering facilities in northern Turkey, probably in the summer of 1955, by means of which the flight-paths and telemetry of Soviet test missiles launched from Kapustin Yar, near Volgograd, could be monitored in considerable detail.[15] The majority of test flights observed were for short-range (170–340 miles) missiles, with some for the original, 500-mile version of the Pobeda. On 9 February 1956 the USAF Chief of Staff, General Nathan Twining, told a closed session of the House Committee on Appropriations that 'we have firm evidence that the Soviets are currently engaged in an extensive ballistic missile development program' and forecast Soviet deployment of missiles with a range of 1600 miles by 1957, 1840 miles by 1957–8, and with intercontinental ranges by 1960–1.[16] But according to the post-sputnik testimony of CIA officials (below) the first Soviet test of a 1000-mile-range IRBM did not take place until June 1957, although the Joint Intelligence Committee (JIC) was aware of 700-mile Soviet missile tests by May 1956.[17] Clearly a longer-range (700–750 miles) version of the Pobeda was being tested by early 1956, and a 1000-mile Soviet IRBM could not be far off, but Twining may have been painting an extreme picture which was not shared by other United States intelligence agencies at the time.

The impact of the technical intelligence from Turkey, and particularly of the first Soviet MRBM tests, was considerable. In February 1956 Secretary of the Air Force Donald Quarles told a New York University audience that 'there is no question of our being ahead of them' in missile development as a whole, and quoted a recent statement from Sir Frederick Brundrett, the Science Adviser to the British Ministry of Defence, to the same effect.[18] On 21 March Quarles told the Aviation Writers Association that 'we would not be surprised to find close competition [from the Soviet Union] in this [ballistic missile] field'.[19] On 16 May the Joint Intelligence Committee noted the testing of a 700-mile range missile, but made no direct reference to the existence or timing of a Soviet ICBM programme as such.[20] On the same day General Goodpaster noted 'the change in the "time table" of Soviet capabilities'.[21] In December Quarles told the press, though with little visible effect, that American priority with a scientific satellite for the IGY was 'by no means assured'.[22]

In November 1957, after the first two sputniks, a closed session of the Preparedness Subcommittee heard a detailed account of the Soviet missile development programme, as it had been revealed to the American monitoring stations, from the Director of Central Intelligence (DCI) Allen Dulles, and a senior aide, Herbert Scoville. An indigenous Soviet missile programme had begun by 1948 and continued steadily ever since. There had been at least 300 test flights since mid-1953 at ranges between 80 and 1000 miles. The first 1000-mile test was on 22 June 1957 and was followed by seven more in two months. No 1500 mile IRBM had yet been tested. There had been 'four tests in ICBM and satellite programs', of which two were long-range tests reaching 4000 miles. A feature of the Soviet programme had been the very rapid growth in the frequency of test firings, which had reached a rate of up to twenty-two a month, and even four a day. It was far from certain that Soviet scientists had yet developed the necessary guidance and re-entry technologies for ICBMs, but deployment of operational Soviet ICBMs was estimated as likely to occur within the 1958–60 period.[23]

That was a significant advance on the dates which had been postulated by Twining just under two years earlier, and in the TCP Report before that. However a more pessimistic estimate than Twining's had probably been communicated to the President and his senior colleagues as soon as it was reached in May 1956. Since the administration had already virtually anticipated it by deciding in December 1955 to add a top-priority IRBM programme to the American effort it is hard to confirm this from policy developments. But in general the contrast with the previous administration's indifference to long-range ballistic missiles was considerable.


## 2 SATELLITE INTELLIGENCE

The Eisenhower administration was both seriously concerned and increasingly well-informed about the Soviet R&D programme for large rockets for ballistic missiles, a programme which in 1951 had begun testing rocket engines with over twice the thrust of that which would eventually provide the first-stage motor for the first American satellite launcher in 1958. In 1954 and 1955, respectively, Eisenhower's government also became committed to an Air Force project to develop an intelligence satellite and to the IGY Committee's scientific satellite project. The latter was periodically reviewed at the highest level of government, the National Security Council. Owing to deletions in the Gleason Minutes it is impossible to be certain, but it seems probable that the former was also monitored at the NSC or at less formal 'kitchen cabinet' meetings. In that context it would indeed have been bizarre if the administration had never asked for intelligence about Soviet satellite-launching capabilities, or had ignored that intelligence when it was provided. In fact, however, the still incomplete public record suggests that such intelligence was requested and provided, though the exact timing is still difficult to reconstruct.

To start once again from the open literature, in the early years of the Eisenhower administration both the specialist and the general press were quick to notice such rare public Soviet statements as the remarks of the

president of the Soviet Academy of Sciences, Alexander Nesmeyanov, in
an address to the World Peace Council in Vienna, that both satellites and
manned lunar exploration were now feasible.[24] At this time Sutton was
also stating that:

> The Russians apparently have considered using a number of these giant
> engines clustered together to power a spaceship. . . .
>   I do not know how far along the Soviets are with a complete satellite
> vehicle. I know that they were planning to work on one several years
> ago.[25]

By 1955 it was plain from the announcement of the Commission on
space exploration and from Sedov's remarks at Copenhagen, however
misinterpreted, that the Soviet Union was interested in the satellite idea
and thinking about the IGY time-frame. The problem for the historian is
to determine the rate at which, after this point, the United States acquired
any harder information about Soviet plans and capabilities in respect of
satellites.

*                    *                    *

The American IGY scientists were usually willing to contribute discreetly
to the intelligence picture. In March 1957, for example, the IGY Committee
secretary Hugh Odishaw asked all American IGY participants to send him
details of any Soviet scientific texts that they might have obtained through
their IGY activities, requiring a reply within a fortnight. A list of such
documents was then circulated.[26]

   On the satellite issue, a confidential report on the Rockets and Satellites
Working Group at the Brussels CSAGI meeting of September 1955 was
prepared for Odishaw. It paid close attention to the level of Soviet
involvement.[27] In the next few months Committee members assembled
a dossier of published Soviet material on rockets and satellites. This was
circulated six months *before* the first similar collection edited by Firmin
Krieger at RAND.[28]

   In February 1956 Hugh Odishaw wrote to Lloyd Berkner enclosing
'the most recent story on possible Russian satellite plans', presumably
the NAS dossier. Odishaw suggested that since not only the US IGY
Committee but those of other countries were keen to know more, the
Secretary General of CSAGI (Nicolet) should be prompted to write
to the Soviet Academy of Sciences to that effect.[29] At other times
the IGY Coordinator, Sir Archibald Day, was on the receiving end of
similar advice.[30]

The IGY scientists were not operating in a vacuum, for there were several channels through which information or informed speculation about Soviet satellite projects could be exchanged with government officials, including the NSF and a Working Group of the NSC's secret Operations Coordinating Board which will be described in the next chapter. James Baker, for example, the astronomer who jointly developed the Baker-Nunn camera for the Smithsonian's optical satellite-tracking programme for the IGY, was a member of the TCP's intelligence committee. It is even possible that the finding of the Joint Intelligence Committee, in May 1956, that 'recent statements by Soviet officials and scientists indicate a high degree of interest in earth satellites and an intention to launch one or more vehicles during the International Geophysical Year, 1957–1958', was based on the Academy's recent dossier.[31]

\*               \*               \*

On the covert intelligence side, the earliest point at which the administration *may* have requested an estimate of the date by which the Soviet Union could launch a satellite was the winter of 1953–4, and the estimate allegedly provided was the second half of 1957.[32] The source of this report, Stewart Alsop, had good contacts in the intelligence community, but supplied no confirmatory details. The story is at least plausible, in view of the possibility, noted in Chapter 9, that Eisenhower was briefed on the satellite idea as early as the autumn of 1953 by the Columbia University physicist John Dunning. The White House could also have picked up on the Nesmeyanov statement (above) or simply on the continuing publicity campaign being waged by von Braun and others, which sometimes moved outside the specialist rocketry journals.[33] Alternatively, such an intelligence estimate could have been sought by the Air Force, which was then in the process of accepting RAND's recommendation for a reconnaissance satellite project.

On the other hand Alsop may perhaps have recalled this event as having come one year earlier than it actually did. It seems rather more likely to have occurred in early 1955, when NSC 5520 shows that the administration was discussing the possibility of a Soviet satellite programme while taking the decision to proceed with an American one in the context of the IGY.[34] The timing of the White House announcement on 29 July 1955 was even changed at the last minute, out of concern lest a similar Soviet announcement deprive the United States of the propaganda advantage.[35] This was probably not due to any specific intelligence about the status of a Soviet satellite project or about the intentions of the Soviet observers who were on the eve of attending a congress of the International Astronautical Federation for the first time. But it did reflect a general American assessment that the Russians were capable of initiating a satellite project and likely to do so in the near future.

The effect of the inaccurate press reports of Sedov's remarks at Copenhagen in early August 1955 on American perceptions of Soviet intentions and capabilities in respect of satellites has already been discussed in Chapter 8. Perhaps the key example of this occurred in a memorandum written by the Assistant Chief of Army Ordnance on 15 August 1955, which gave as one of the arguments for reversing the Stewart Committee's selection of the Navy's Vanguard satellite-launching system, and substituting one based on an improved Redstone missile, the consideration that:

> The first orbital flight for this configuration can be scheduled for January 1957 if an immediate approval is granted. Since this is the date by which the U.S.S.R. may well be ready to launch, U.S. prestige dictates that every effort should be made to launch the first U.S. satellite at that time.[36]

Between 1955 and 1957 the National Security Council received regular and fairly frequent briefings from DCI Allen Dulles on the scale and pace of the Soviet missile programme. The degree to which such presentations also dealt with the question of artificial satellites is unclear, since the relevant passages in the Gleason Minutes have not yet been declassified. The state of the intelligence picture during 1956, which was probably the crucial period for American decision-makers, is particularly obscure. The extended NSC discussion of the satellite project on 3 May 1956 touched obliquely on the continuing efforts of the Army Ballistic Missile Agency to obtain permission for a satellite launch-attempt with its Redstone-based system, Juno 1, when 'Secretary Wilson interrupted . . . to indicate that when one got into the area of alternative launching vehicles such as Redstone, there was real danger of a conflict between the earth satellite program and the ballistic missile programs of the Defense Department'.[37] Part of the rationale for the ABMA proposal was of course that such a shot could be made in the first half of 1957, with a view to a putative Soviet ability to do likewise. There is a deleted paragraph at the end of the Gleason Minutes of the discussion, but no reference was made at any other point to the Soviet side of the equation.

A few days later there may have been a significant change. The passage already quoted from the JIC's estimate on 16 May refers only to a possible Soviet launch attempt within the IGY. But Goodpaster's remark about 'the change in the "time table" of Soviet capabilities', which was almost certainly a comment on the JIC paper, may possibly have referred rather more to satellites than to missiles. The formal title of the committee of scientists chaired by Homer Stewart which was advising the Defense Department on satellites, it will be recalled, was the Advisory Group on Special Capabilities. Soon after it was established those involved began using the word 'capabilities' as an informal code for 'satellites'.[38] If Goodpaster picked up this idiom, as he could have done, the new timetable he was noting on 16 May 1956 may have been that estimated for a Soviet satellite project.

The Gleason Minutes suggest two possible occasions in 1957 on which Dulles may have briefed the NSC about the state of the Soviet satellite programme, on 24 January and 10 May.[39] After Sputnik 1 Dulles himself stated that 'our own intelligence in analyzing the Soviet emphasis in the field of rocket propulsion and its announced program gave us *last March* an estimate that the Soviet could in fact orbit an earth satellite in 1957, as they have just done'.[40] But in memoirs apparently based on documentary sources Eisenhower himself gave November 1956 as the date for the first estimate that the Russians would be able to launch a satellite any time after November 1957.[41] Lastly, the former CIA analyst Ray Cline has recalled briefing the NSC on Soviet satellite preparations at some point *after* 'mid-year' 1957. His original account seemed to place this NSC discussion rather close to the launching of Sputnik 1, but he has since suggested it was some months earlier than that.[42]

\*          \*          \*

Most of this admittedly scanty evidence becomes even less convincing upon review. Both Alsop's story and the conjecture which has been put forward about Goodpaster's remark in May 1956 are shaken by two considerations. The first is that none of the declassified intelligence estimates from 1954 to mid-1956 contains anything more definite about a future Soviet satellite shot than that it would probably be made at some point within the IGY, in other words between 1 July 1957 and 31 December 1958. There is no suggestion in the 1956 JIC paper, for example, that a Soviet launch attempt might come *early* in the IGY. The second is that in October 1957 Allen Dulles himself only recalled that an intelligence estimate that a Soviet launch attempt might come during 1957, that is, during the first six months of the eighteen-month IGY, had apparently *first* been made in March 1957.[43] For 1956, it is also interesting that even the papers on the Redstone option which were submitted to Deputy Secretary of Defense Reuben Robertson in July 1956 made no direct reference to the possibility of an early Soviet launch.[44]

The apparent difference of five months between the recollections of Eisenhower and Dulles, about when exactly they acquired intelligence that the Soviet Union was well on the way to launching a satellite, may be unimportant. Though both refer to a Soviet capability rather than an actual project, Dulles may have based his date, March 1957, on a firmer quality of intelligence than the material which Eisenhower later recalled. Until the document referred to by Eisenhower comes to hand the balance of credibility should lie with Dulles, who was writing closer to the event and was the more expert witness of the two. The news was passed to the NSC and other concerned agencies between then and June, when RAND circulated its prediction that the Russians would launch on about 17 September, the centenary of the birth of Konstantin Tsiolkovsky, the Russian pioneer of rocket theory.[45] On 12 September General Gavin told the Army Scientific Advisory Panel he thought there would be a Soviet launch within thirty days.[46] A less focused estimate may also have been repeated to the NSC at the eleventh hour.

*     *     *

Finally, it is worth comparing the performance of the American intelligence services with what little is now known about the two key Soviet decisions which would have been their targets. The first was the decision to proceed with a definite IGY satellite project – the Soviet equivalent of approving NSC 5520. A recent Western account based on Soviet sources states that 'the final go-ahead was given at a closed meeting held in the Academy of Sciences in Moscow on 30 August 1955'.[47] In another version the senior Soviet rocket engineer and satellite project director, Sergei Korolev, is said to have recalled that the final decision to proceed was reached in January 1956.[48] In view of the limitations of the Soviet bureaucracy of the day, either date would explain why Soviet scientists were not ready to make an announcement to the Brussels CSAGI meeting in September 1955, but put off doing so until the Barcelona meeting a year later. The choice of a time-frame within which to attempt a launch, the second decision that would have been targeted by American intelligence, would probably have been made in general terms before June 1957, when the first article describing the satellite and its radio-tracking system was published in *Radio*, but a more precise date is unlikely to have been settled until after the first successful full-range test of the ICBM configuration of the launching rocket, which took place on 3 August.[49] On 18 September Radio Moscow confirmed General Gavin's recent prediction by telling the people of the Soviet Union that a launch was imminent.[50]

# 11 Eisenhower's Early Space Policy

## 1 APPEARANCE AND REALITY

The early space policy of the Eisenhower administration was made up of several elements. To identify them it is necessary to look carefully at everything that the administration actually did and to avoid mistaking what it said for the whole of its policy. It is even more necessary not to confuse what was said by Eisenhower and his colleagues *after* the first sputniks with their actual space policy before that point.

The strong interest of the Eisenhower administration in strategic reconnaissance has already been described in Chapter 9. Among its fruits were the DEW Line radar system for early warning of strategic attack, the radars and listening posts for technical intelligence-gathering in Turkey and several other countries, and the crash programme to develop the U-2 spy plane and fly it against Soviet and other strategic targets in the second half of the 1950s. The RAND Corporation persuaded the Air Force to adopt its recommendation for a full-scale reconnaissance satellite project in 1953 and 1954, Eisenhower's first two years in office, and the available evidence suggests that by the autumn of 1954 this activity was known to and supported by the President and his closest colleagues.[1] However, in the judgement of the administration's scientific and political advisers there were considerable scientific and diplomatic difficulties to be overcome before reconnaissance satellites could make a reliable contribution to the security of the United States. There was also the mundane but relevant consideration that launching systems powerful enough to place in orbit the size of payload envisaged for reconnaissance satellites would not be available until sufficient progress had been made with the Atlas ICBM project, for which the development of an operational weapon was in any case the top priority. Although the Eisenhower administration had hardly been responsible for the late start on ICBMs it could not escape the consequences. Meanwhile smaller rocket launchers, capable of orbiting small but scientifically useful payloads, were becoming available.

The IGY satellite proposal promoted by the National Academy of Sciences had the advantage of promising to help with both types of difficulty envisaged for future reconnaissance satellites, by providing both the necessary advances in scientific knowledge and technical experience and a climate of positive international attitudes towards activities in space. Even if the early scientific satellites could not actually pass on their predicted international legitimacy to their military successors, they would at least cloud the issue for a short time, and provide a suitable cover story for even longer, as in fact happened in 1959–61 when the CIA's Corona programme to develop a photo-reconnaissance satellite with a recoverable capsule was conducted under the cover of the Air Force's Discoverer programme of experimental 'biosatellites'.[2] By mid-May 1955 the RDB's Coordinating Committee for General Sciences had reported favourably to Quarles on the idea of a satellite and a positive military evaluation of the project had also been prepared, but no decision had yet been reached about a launch system.[3] A policy paper endorsing a scientific satellite project for the IGY made its way through the relevant departments and was ratified by the National Security Council on 26 May. It laid down some important reservations. The project might be placed 'under international auspices, such as the International Geophysical Year, in order to emphasize its peaceful purposes'. But:

. . . such international auspices [should be] arranged in a manner which:
a. Preserves U.S. freedom of action in the field of satellites and related programs.
b. Does not delay or otherwise impede the U.S. satellite program and related research and development programs.
c. Protects the security of U.S. classified information regarding such matters as the means of launching a scientific satellite.
d. Does not involve actions which imply a requirement for prior consent by any nation over which the satellite might pass in its orbit, and thereby does not jeopardize the concept of 'Freedom of Space'.[4]

Of these conditions the most interesting is perhaps the second, since it reveals that as far as the NSC was concerned the United States already had a satellite programme that was entirely distinct from the IGY satellite project discussed in the document.

NSC 5520 also made it clear that, however peaceful its purposes, the IGY satellite's rocket launcher would have to 'be developed from existing missile programs within the Department of Defense'. Precisely because of this connection, care should be taken to ensure that the project would not 'materially delay other major Defense programs', presumably a reference to missiles.[5] When the project was made public on 29 July 1955 no attempt was made to disguise the fact that 'the required equipment and facilities for launching the satellite' were to be provided by the military.[6] Subsequent statements continued to point this out.[7] It was also made clear at the time of the original announcement and thereafter that although 'the scientific satellite . . . is not primarily a military device or primarily directed toward military information . . . it will, of course, broaden our base for a great many things, including military technology'.[8]

Far from setting an unreasonable requirement that the IGY satellite project should somehow be *actually* 'non-military' or 'civilian', as many historians have seen it, the Eisenhower administration's policy before the sputniks was complex and sophisticated.[9] When the General Sciences Committee played safe by supporting the satellite proposals of all three services and proposing a six-months postponement of the decision, Quarles set up an eight-member Ad Hoc Advisory Group on Special Capabilities (the Stewart Committee) with instructions to select a launch system for the IGY satellite by 1 August. During June and July the Stewart Committee made a careful survey of the alternative proposals and evaluated each against a list of nine criteria. The requirement that the system selected should occasion minimum delay to military projects was placed sixth. Above it came the predicted launch capacity of the system, the growth potential of the selected design, and the best use of available facilities and skills. Immediately after the non-interference criterion came the requirement that the system chosen should have maximum scientific utility.[10] However the non-interference criterion really only served to rule out an unrealistic Air Force offer of the still embryonic Atlas ICBM booster, a point referred to in the Committee's final recommendation, which mentioned Vanguard's lack of interference 'from or with weapon projects', but only after citing its potentially better performance and fewer rocket stages.[11] The choice between the two remaining contenders, the Army's Redstone-based Orbiter and the Naval Research Laboratory's Viking-based Vanguard, which Clifford Furnas later recalled as having been a 'toss-up',[12] could not be resolved so easily. The point was that, although the Redstone was a missile and the Viking had been primarily an atmospheric research vehicle, albeit with overtones of the Navy's interest in acquiring its own long-range ballistic missiles,[13] the Army's Redstone Arsenal team under von Braun had virtually completed the development work for the Redstone, which was about to be handed over to Chrysler for the production phase, and was not assigned to develop the Jupiter IRBM until some months after the selection process for a satellite launcher was completed.[14]

\*                    \*                    \*

On 3 August 1955 the Stewart Committee decided to recommend a version of the Navy's Vanguard proposal, with the chairman and Furnas dissenting. Quarles's R&D Policy Council discussed their report three days later but postponed a decision on its findings. The Army succeeded in forcing a second round of presentations at the Stewart Committee, from which the Navy took care not to be left out. The balance of the Committee's judgement was unaltered, and the Policy Council endorsed it at the end of August with the Army still dissenting. In view of later charges that the decision process had been premissed upon indifference to political considerations, it is interesting to note Quarles's later testimony that 'there was consideration given to [the impact which a Russian victory in this field would have on world opinion], and I would say that that was certainly one of the reasons it was authorized in the first place'.[15]

During 1956, as the Soviet satellite programme was increasingly realized to be a serious competitor, the Army continued to press the merits of Orbiter as a potentially rapid achievement and prepared several copies of the relevant 'Jupiter C' configuration, Redstone with solid-fuel upper-stages, for long-range warhead re-entry tests. In June 1956 the Stewart Committee was consulted on the question of whether an early Orbiter-type satellite shot should be attempted in order to compensate for the delays arising in the Vanguard project. It found that such a shot could be made at the end of 1956 or early in 1957, but with a less than 50 per cent probability of success. It also pointed out that such a launch would be too early for the IGY and claimed, erroneously, that no ground-tracking facilities would be in place to support it, so that it would fail to fulfil the American commitment:

> Moreover, any announcement of such a flight (or worse, any leakage of information if no prior announcement were made) would seriously compromise the strong moral position internationally which the United States holds due to its past frank and open acts and announcements as regards VANGUARD.[16]

This position was strongly supported by Clifford Furnas, who had succeeded Quarles as Assistant Secretary for R&D. He warned that any panic switch to Orbiter would damage the morale of the Vanguard team and disrupt 'our relations with the non-military scientific community and international elements of the IGY group'.[17] This view was passed to Eisenhower by General Goodpaster, who noted on his copy that '[Deputy] Secy [of Defense] Robertson feels no change should be made.'

On 20 September 1956 the von Braun team sent the dummy fourth stage of a Jupiter C test shot a distance of 3335 miles through an apogee of 682 miles.[18] After the sputniks ABMA commander General Medaris and others made much of this undoubted technical achievement, claiming that if the upper stage had been live it could have launched the world's first satellite there and then, one year ahead of Sputnik 1, or that this could at least have been done four months later – the time which was in fact taken after 4 October 1957 to prepare a successful Army launch. But the Army's actual submission in October 1956 was that they could launch by June 1957.[19] The tracking considerations brought up by the Stewart Committee were still relevant. It may also have been noticed that even the Army's timetable was starting to slip, doubtless because since November 1955 their rocket team had been assigned to develop the Jupiter IRBM. In short the non-interference consideration now counted against the Army, whatever the difficulties being encountered by Vanguard.

*          *          *

Unfortunately, the historical perception of this aspect of Eisenhower's early space policy has been coloured by the fact that after Sputnik 1 the administration staged a damage-limiting public-relations exercise, in line with its previous declaratory policy, which drew a bogus but meretricious distinction between its own peaceful, scientific and allegedly disinterested satellite project and the somehow more sinister and less noble, if more effective, Soviet one. Since it was only at this point that the administration began to emphasize the separation of Project Vanguard from missile programmes, it is of some interest to discover the origins of this apologetic policy.

Both the DCI, Allen Dulles, and the Secretary of State, John Foster Dulles, drafted possible presidential statements over the weekend after Sputnik 1, but neither came up with the idea of stressing the 'non-missile' character of the Vanguard rocket. That approach had originally been conceived over three months before 4 October 1957 by a secret Working Group on Project Vanguard which had been set up by the Operations Coordinating Board in July 1955, at the time of the decision to proceed with an IGY satellite project. In June 1957 the Working Group responded to a short-lived scare about a possible Soviet launch in the summer by deciding that:

> . . . each department and agency would take the necessary precaution-
> ary measures from a public relations standpoint to insure that (a) the
> United States disclaim any intention of engaging in a race with the

Soviets to launch the first satellite; (b) that commentary by leading U.S. scientific officials be developed on a standby basis which would accept the Soviet development in the cooperative spirit of the International Geophysical Year.[20]

The minute shows clearly enough that this was not intended to be the *de facto* or 'action' policy of the United States, but simply a pre-prepared public-relations response to the potential Soviet achievement. When June passed without a Soviet triumph the statement 'proposed for use by Dr Detlev W. Bronk, President, NAS' was shelved.[21]

When the Working Group convened on 5 and 7 October 1957 in line with its responsibility to 'concern itself with the impact of Soviet Earth Satellite programs', the June contingency statement was naturally cited as the 'pattern' which all departments were now to follow. Statements should 'avoid any reference to the military implications', 'congratulat[e] the Soviet scientists on their accomplishments', and 'Play down competitive aspects and implication of a "race" in development and launching'.[22] The person who then sold this public relations approach to Eisenhower was OCB member Donald Quarles.

Nothing better illustrates the fact that the Working Group's recommendation to 'Keep the accomplishment within a peaceful context' was only meant for the administration's public posture than the alarmist anti-Soviet memorandum which Quarles wrote on 7 October and presented to the President the following day. In it he referred to the 'independent course of rocketry' followed by the American project. After noting that it had been understood all along that 'defense participation would be to supply the rocketry needed to place such a satellite in orbit', he added that 'the military participation in rocketry was *de-emphasized* as being incidental to the scientific program. All subsequent public releases have *followed this same line*.' That hardly did justice to the fact that the scientific satellites and their instrumentation were being designed and built by a consortium of civilian and military institutions, coordinated by the Naval Research Laboratory, but the point of course was that Quarles was recapitulating the 'line' which had been followed in public information, not the whole truth about the American project. By contrast, 'the Russian satellite work has been closely integrated with and has drawn heavily on their ballistic missile developments'. The record of Soviet actions allegedly showed that 'a very carefully laid plan to make maximum cold war capital out of their satellite program' had been initiated as early as the Rome CSAGI meeting of October 1954. Supposedly the Russians had been able to give their satellite equal priority with their missile programme because they knew they had never been under any threat of attack by the United States, whereas the American satellite project had been unfairly handicapped because the converse was not the case.[23]

Eisenhower wisely forbore accepting all of Quarles's points, but he did agree to run with the excuse that 'The U.S. determined to make the Satellite a scientific project and to keep it free from military weaponry to the greatest extent possible.'[24]

\*                    \*                    \*

The question of whether or not the separation of the IGY satellite project from the American missile program was part of a disinterested policy of support for 'pure science' can be further investigated by looking at the records of the project's secret political management.

## 2    THE OCB WORKING GROUP

The Eisenhower administration was concerned from the start that the IGY satellite project should be conducted and publicly presented in such a way as to secure the maximum benefit for the prestige and influence of the United States in the propaganda competition with the Soviet Union. The government organization with primary responsibility for such aspects of American policy was the Operations Coordinating Board (OCB), which had been established by Eisenhower on 2 September 1953 with Under Secretary of State Bedell Smith as its first chairman. Other founding members were Deputy Secretary of Defense Roger Keyes and DCI Dulles.[25] A key role was played by C. D. Jackson, Eisenhower's Special Assistant for International Affairs, or as he was informally known, 'for cold-war strategy'. After Jackson left the White House to return to publishing in March 1954 he was succeeded as Special Assistant by Nelson Rockefeller. But at the time of the satellite decision in 1955 the leading role at the Board appears to have been played by its executive secretary, Elmer Staats.

The OCB replaced Truman's Psychological Strategy Board (PSB), which was criticized as ineffective by Jackson and others during Eisenhower's first months in office. Among other things it was claimed that the PSB had not even made contingency plans for American responses to Stalin's eminently foreseeable death.[26] In its own view the OCB served 'as the body for dealing with requirements for interdepartmental coordination concerning overseas information and psychological warfare activities in carrying out NSC assignments or upon specific request by participating departments and agencies'.[27]

The Board was purely consultative, with no executive powers. At first it was an independent organ within the White House with some responsibility to the State Department. But in February 1957 Robert Cutler was appointed to direct the NSC as Special Assistant for National Security Affairs and also became OCB Vice-Chairman for a few months. At this point it was decided to bring the Board within the NSC structure from July of that year.[28] Another strong personality, Frederick Dearborn, served as Special Assistant for International Affairs from May 1957 until his death in February 1958.

*          *          *

NSC 5520 pointed out that 'Considerable prestige and psychological benefits will accrue to the nation which first is successful in launching a satellite', and saw it as highly desirable to prevent the Soviet Union from achieving that distinction. The OCB had already been seised of the matter, and Nelson Rockefeller underlined both points in an appendix.[29] Within two weeks of the NSC's approval of the project NSF director Alan Waterman was discussing its implementation with the OCB's Elmer Staats. Waterman emphasized that 'the Foundation has the responsibility for administering the IGY program', an arrangement which 'insures that the scientific planning for the IGY is properly coordinated in one government agency'.[30] By 22 July an 'Ad Hoc Working Group on Information Aspects of NSC 5520' was constituted. Its membership included Waterman and J. Wallace Joyce, the head of the NSF's IGY office, together with representatives from the Departments of State and Defense, the CIA, OCB and US Information Agency.[31]

The Working Group was mainly concerned, at first, with ensuring that the announcement of an IGY satellite project would give maximum benefits to the United States. A letter to Sydney Chapman, president of CSAGI, in the name of Joseph Kaplan, chairman of the officially non-governmental US IGY Committee, was drafted and re-drafted. Kaplan objected to this procedure on the grounds that he had no right to take such a step without consulting his committee, but he was powerless to prevent it.[32] On 22 July Berkner wrote to Nicolet informing him that something so important was about to be announced that an official of the National Science Foundation would wait on him in person with the details in a week's time.[33] A copy of the final version of 'Kaplan's' letter, dated 26 July, was duly delivered to Nicolet for release on the 29th, in what Eisenhower had ordered to be a carefully orchestrated conjunction with the White House announcement.[34]

To complement 'Kaplan's' letter the Working Group also drafted a press release in the joint names of the National Science Foundation and the National Academy of Sciences.[35] Waterman and Under Secretary of State Herbert Hoover then presented their public information programme to Eisenhower, Quarles and Goodpaster at the White House.[36] At a Working Group meeting on 3 August members 'expressed satisfaction with the fact that civilian scientific aspects of the satellite were emphasized and the military connotations were played down'.[37] On the same day Kaplan wrote to Waterman with a good grace to express his satisfaction that 'in the course of the announcement of the program and subsequent news enquiries . . . much good use was made of the material prepared by our Committee'.[38]

After Sputnik 1 the original satellite announcement was recalled by commentators as 'an important propaganda victory' in itself.[39] But in retrospect the activities of the Working Group exemplify an aspect of the Eisenhower administration described by one historian as 'faith in the efficacy of public posture: the belief that by merely making pronouncements and striking public poses, the United States could increase the difficulties under which its opponents operated'.[40]

*          *          *

After 8 August the Working Group was suspended for a time, and the satellite project was therefore omitted from a list of OCB activities drawn up in September.[41] On 12 October the Defense Department stepped in to convene an *ad hoc* inter-agency liaison group on the satellite project which for a time took over the role of coordinating public information.[42] It also dealt with such matters as the problem of obtaining launch facilities for Project Vanguard at the Air Force Missile Test Center at Patrick Air Force Base in Florida (Cape Canaveral), progress in negotiations with the Army about satellite observation stations, and the difficulty of channelling funds for Vanguard to the Naval Research Laboratory.[43] After toying with the idea of responding to this bureaucratic challenge by creating an over-arching Working Group 'On Coordination of IGY Programs and Related Scientific Programs' the OCB simply reconvened its satellite Working Group in January 1956. At the end of February the dispute between the Board and the Department of Defense over the location and terms of reference of a more permanent liaison body was resolved with an agreement to suspend the Pentagon's inter-agency group and reconvene the OCB Working Group under Defense chairmanship.[44]

The reconvened Working Group 'On Certain Aspects of NSC 5520' did not begin meeting until May, after the NSC had ruled on the difficult question of the size of the Vanguard programme.[45] While it continued to discuss public information and propaganda issues, such as arrangements for international witnessing of the satellite launch, other items of business reflected its new, broader remit. At its first meeting on 18 May 1956 it discussed the problem of ensuring international cooperation in tracking the satellite, particularly from Soviet and Communist Chinese ground stations. On 28 August it heard a report from the Stewart Committee to the effect that 'all program commitments were on schedule and that elements of skepticism with respect to the success of the program were diminishing'.[46] There were no further meetings until those of June 1957, at which the Working Group devised the response to a possible Soviet announcement described above. The point perhaps bears repeating that far from the administration's political leaders having had to cast around after Sputnik 1 for the excuse that the IGY satellite project had not been a 'race' against the Soviet Union, this approach was prepared some months in advance by a group of middle-ranking officials, military officers and scientists, at some remove from the Oval Office.

## 3   THE ANTARCTIC PRECEDENT

The events just sketched show that the satellite project of the purportedly non-governmental National Academy and its IGY Committee was assumed from the outset to fall within the purview of the administration's cold-war apparatus. If conscientious scientists such as Kaplan offered only token resistance to such interference, that was perhaps because the satellite project was not the first part of the American IGY programme to be subjected to it. The minutes of earlier meetings of the National Academy's Antarctic Committee for the IGY show that it too was affected by a special OCB Working Group. Thus in October 1954 the OCB requested and obtained, 'in view of national interest', an immediate IGY presence on board the Navy icebreaker USS *Atka*, which was about to make a reconnaissance trip to the Bay of Whales and the Weddell Sea. That was one year sooner than the scientists had actually intended to begin their field-work.[47]

The OCB further intervened with the Antarctic Committee over the delicate question of which other countries should be encouraged or assisted to set up research bases in areas of potential relevance to future United States claims on the continent. It also ensured that a suitable press release was prepared for use by the IGY Committee in conjunction with the 'presidential statement' on the American IGY expedition to Antarctica that was issued from the White House by Eisenhower's press secretary James Hagerty on 28 March 1955, a pattern that would be closely repeated four months later with the satellite announcement.[48]

Another aspect of relations between the Antarctic Committee and the OCB which was to be repeated in the satellite case was the difficulty which the NAS scientists experienced in bypassing the NSF and obtaining direct access to the relevant OCB Working Group. The Antarctic Committee minutes contain the observation that:

> ... the I.C. [Interdepartmental Committee] is an interdepartmental group and therefore there would be a delicate problem in approach since the IGY [*sic*] is not technically a government agency. Nevertheless. the Committee pressed this point and the desirability of immediate stronger liaison was urged.[49]

## 4 THE ACADEMY AND THE WORKING GROUP

In the satellite case the bureaucratic difficulty was exacerbated by a general tension between the National Academy of Sciences and the National Science Foundation over the question of which organization was ultimately responsible for coordinating the IGY programme in general and the satellite project in particular. At first, as we have seen, only the National Science Foundation was represented at the OCB Working Group. By December 1955 Joyce was working with Hirsch of the OCB on the first of many versions of a set of guidelines for information management throughout the IGY programme, including the satellite project.[50] Apart from limited cooperation with the US Information Agency, which was represented at the OCB, the Academy's IGY scientists were not involved in this process.

Hugh Odishaw, the secretary of the US IGY Committee, had attended both the October 1954 and the January 1955 meetings of the Antarctic Committee, and its chairman Joseph Kaplan had attended the second of these. By 28 July 1955, when unnamed members of the IGY Committee attended a meeting of the OCB's satellite Working Group, both these Committee officers were familiar with the pattern of OCB intervention in areas of proposed IGY activity that were deemed politically sensitive by the administration.[51]

*                    *                    *

Early in 1956 the scientists working on the IGY satellite began to have problems in the very area of public relations for which the OCB had made itself primarily responsible. In February the Eisenhower administration had been obliged to respond to vigorous Soviet protests by terminating a US Air Force project for gathering photographic intelligence of the Soviet Union by means of high-altitude balloons released from sites in Western Europe, code-named Project Genetrix. Balloon releases began on 10 January 1956 and were discontinued on 6 February, after 387 balloons had crossed the frontiers of the Soviet Union or its allies. A cover story that the balloons were for meteorological research was supported by a few releases in Okinawa, Hawaii and Alaska, the first of which was staged with international press coverage on 9 January.[52] In response to the Soviet objection, an Air Force press statement on 7 February claimed that 'This method of obtaining meteorological data might be of great use in the International Geophysical Year program that will be conducted during 1957–58.'[53]

Loyal Western news reports and editorial comment proceeded to inflate the Air Force's carefully planted suggestion into a certainty, much to the chagrin of the US IGY Committee, whose members feared that 'this linkage could seriously affect the cooperation of other nations in the IGY program'. Their objections having been ignored by the relevant government agencies, the scientists concluded optimistically that there was a need for 'direct relationships at appropriate policy levels' and for advance review by themselves of *all* government statements referring to the IGY.[54]

Of more direct concern for scientists working on the IGY satellite project were the press reports which began to appear in the first half of 1956 stating that Vanguard was falling seriously behind schedule.[55] At the end of April Armand Spitz, a consultant working with the Smithsonian Astrophysical Observatory's Moonwatch programme for training amateur satellite observation teams, gave a press conference in Montreal from which he was reported as saying that the Russians were likely to be first with a satellite.[56] This was regarded as such a public-relations catastrophe that an official of the IGY Committee initiated an internal disciplinary process on the grounds that 'the content of the above news report is contrary to the interests of the USNC-IGY'.[57] Spitz later explained to Odishaw that he had replied to a question about the timing of a Soviet launch as follows:

I had read the A. P. [Associated Press] dispatch following Eisenhower's July 29 announcement, saying that the Soviet planned to launch a satellite before we launch ours. I said that this is all I know, that *there seems to be no reason to question their ability to do it*, but that our concern should be directed to planning for observation of our own satellite, – and possibly, in this preparation, we might develop the ability to find other satellites than ours.[58]

If the 'it' of which Spitz had conceded the Russians were capable was launching first, and not simply launching, his complaint of having been misrepresented by the press was perhaps excessive. But according to the then director of the Observatory, Fred Whipple, this was not an isolated incident, since several attempts were made to control the scientists' public statements for propaganda purposes. The staff of the US IGY Committee had allowed themselves to be directed by 'the secret people' into an 'unbelievably stupid' policy of muzzling public discussion of the Soviet satellite project, because 'they didn't want anybody to believe the Russians could do anything'.[59]

*                    *                    *

In early June 1956 Kaplan, Odishaw and Waterman had an important discussion over lunch at the Pentagon with Clifford Furnas, a former member of the Stewart Committee, Quarles's successor as Assistant Secretary of Defense for R&D, and current chairman of the OCB Working Group, and Admiral Rawson Bennett, head of the Office of Naval Research within which the Naval Research Laboratory was responsible for developing the Vanguard satellite-launcher. The primary purpose of the meeting was to sort out Vanguard's obstructed finances but the group also reviewed the project's public information procedures. Kaplan had not given up on the struggle for overall control to be vested in the National Academy of Sciences. According to him, Furnas and Bennett both 'agreed to the language of our proposal' under which the National Academy of Sciences would have sole responsibility for news releases on the satellite, though the OCB would be entitled to 'review' any information released which originated from a government agency.[60] Kaplan rather naïvely interpreted Waterman's silence at this point of the discussion to mean that he was accepting the minimal role which the document had assigned to the National Science Foundation, which was solely that of being 'fiscally responsible to the Government for proper accounting for IGY funds expended in the program under recommendations of the Academy', while the lion's share was bluntly assigned to the latter:

2. Definition of Responsibilities
2.1 The National Academy of Sciences (NAS) has over-all responsibilities for the planning, direction, and execution of the satellite program and specifically for (i) all aspects of the scientific program . . . and (ii)

the international and related national relation [*sic*] problems inherent in the total IGY effort. . . .

4. National and International Public Information

The Academy has primary over-all responsibilities for the dissemination of technical and public information. Academy representatives are to be included in groups concerned with Federal policy in this as well as other areas involving the program.

a. Releases incorporating significant new information, broadly involving national or international commitments of the Academy shall be issued by the Academy with consultation of the inter-departmental group which shall review all such information having its source in the Government.

b. Public information releases involving routine technical matters may be released by the several institutions where it clearly does not involve subjects that fall within the area covered by item (a) above . . . [61]

In fact Waterman seems merely to have been using the Academy in his struggle with the Pentagon for control over the political side of the satellite project, and was far from ready to allow the Academy *carte blanche* in aspects of public relations 'which may have serious international implications'.[62] Another member of the IGY Committee, Richard Porter, took part in a similar meeting with Bennett, Furnas and Waterman at about the same time as Kaplan's, and recorded an informal agreement that 'The Earth-satellite program should be thought of as an IGY project in which the Department of Defense is participating, rather than as a D.O.D. project.' But when it came to the control of public information the group could only agree that the Pentagon should have the final say on 'security' aspects and the Academy on 'policy', while the State Department would also have to be consulted.[63]

The bureaucratic confusion continued, with Kaplan conceding at one point that there could sometimes be joint Academy-Foundation releases, but insisting that it was up to the Academy, not the Foundation, to determine when this should happen. Otherwise, he told Waterman, 'I foresee grave danger if the Committee . . . permits any discipline to become cloaked in governmental authority or even the appearance of governmental authority.'[64]

But the Academy's defence of its scientific independence was perhaps not all that it seemed. The IGY Committee may have reopened this area of friction with the Pentagon and the National Science Foundation as a move aimed at securing full membership of the OCB Working Group, from which Waterman had excluded them on the grounds that they were not a government agency. And indeed Odishaw was eventually admitted to regular membership of the OCB Working Group from its meeting on 28 August 1956.[65]


## 5   THE WORKING GROUP BEFORE SPUTNIK 1

Despite gaining access to the government's policy and propaganda apparatus for the satellite project, the scientists continued to experience problems in this area during 1956. The most serious was the tendency of various agencies to indulge in over-confident forecasts of the likely outcome. In the letter which he wrote as chairman of the ODM Science Advisory Committee to the ODM's director Arthur Flemming in October 1956, to support the IGY Committee's request for additional Vanguard launch attempts, Isidor Rabi also drew attention to this problem:

> Since the success of this scientific program cannot be guaranteed, it is important that our official announcements take account of this, referring to a 'satellite' only after it has been placed in orbit.[66]

The relevance of such advice was underlined by a press conference at the Naval Research Laboratory in December 1956, which gave what some perceived as a 'virtual guarantee' that the United States would have a satellite in orbit by the end of 1957.[67] Officials meeting at the Working Group called for a black-out on the satellite project throughout the Department of Defense 'pending the final resolution of the guidelines paper on the publicity for the satellite'.[68] The guidelines were duly reissued on 23 January 1957. But this version, which remains classified, seems not to have included the warning added *after* Sputnik 1:

> Releases and statements . . . should . . . avoid arousing unfounded expectations by over-commitment in areas of uncertainty such as launching schedules, and possibilities of achieving orbit.[69]

At the National Security Council in January 1957 no one asked Waterman to explain his carefully-phrased remark that, while the United States was ahead of the Soviet Union 'in the development of *instrumentation* for satellites . . . the Russians *might well surprise us in the achievement of a launching unit*'.[70] But in May, after a review of what Secretary of State

Herter termed the 'rather frightening picture' presented by the project's financial disarray:

> The President then commented that there was one lesson to be learned from the experience with the earth satellite program: In the future let us avoid bragging until we know we have succeeded in accomplishing our objectives.[71]

However, this message never really got through to those who were responsible for issuing public reports on the progress of Vanguard. On 3 June 1957 the Deputy Chief of the Naval Research Laboratory, Captain Alfred Metsger, told the Senate Appropriations Committee that:

> Sometime during the coming year, a small new object is scheduled to appear in the skies above us. . . . It will be the first artificial satellite. This earth satellite will be a 20-inch magnesium sphere . . . [etc.][72]

Even if the time-scale was not unreasonable, the specific description which followed the general statements plainly implied that the 'first artificial satellite' would be the full-scale Vanguard.


## 6  THE WORKING GROUP AFTER SPUTNIK 1

The IGY Committee was represented at only one of the two Working Group meetings held in 1957 before 4 October, but was present on a regular basis between that date and February 1958. Immediately after Sputnik 1 it was noted that official responses to the Soviet achievement had broadly followed the political line established by the Working Group's 'contingency statement' in June, to which the Academy had contributed (the 'Bronk' statement).[73] This had emphasized the virtues of international scientific cooperation and avoided any references to a space race or to Soviet geopolitical gains.

The Working Group then asked the IGY Committee to prepare a contingency statement for use in the event of an anticipated second Soviet satellite launch. This went through at least three versions.[74] The scientists also pledged their political support for the administration's cold-war objectives in the form of a general report on the IGY in which 'special emphasis would be placed on spelling out the nature of the IGY agreements on exchange of information to which the Soviets had *professed adherence*'.[75] In a similar vein, after the launch of Explorer 1 on 31 January 1958 the scientist-propagandists undertook to pursue 'the psychological advantage which would accrue to the United States from early report to the IGY World Data Centers of the scientific information obtained from Explorer'.[76]

Another political decision to which the scientists contributed at the OCB concerned the appropriate response to a Soviet offer of payload space for American instruments on a future Soviet satellite. Their solution was to offer a 108 Mc/s transmitter 'to place in one of the Soviet satellites so that its radio signals could be monitored by all IGY tracking stations'.[77] In view of the accusation that the Soviet Union had broken an IGY agreement about satellite frequencies this was clearly intended to make a cold-war point.

The Working Group spent some time discussing the psychological-warfare gains that might accrue from naming American satellites after certain individuals. In November 1957 a list of European and American explorers was drawn up with the name of Tenzing added awkwardly at the end as an 'outstanding Asian'; the name 'Explorer 1', which was subsequently given to the first (Army-launched) American satellite by the Secretary of the Army and the Chief of Staff, probably derived from this process.[78] Then the Academy asserted its right to name the American satellites independently of the government, perhaps beginning with a 'Goddard', 'in honor of the father of U.S. rocketry', and moving on to various astronomers, including 'Arab and Asian personages'.[79] (The idea was later dropped in favour of a version of the usual, abstract astronomical designations for short-lived objects.)[80]

When the Soviet Academy of Sciences wrote to the US Academy in December 1957, claiming (in response to a US Army announcement) that fragments of the upper stage of the booster rocket for Sputnik 1 had landed in Alaska and on the west coast of Canada and the United States, the IGY Committee's reply was cleared with the Working Group.[81] Eventually a paper rejecting the Soviet claim was presented at the Moscow CSAGI meeting.[82]

Despite the IGY Committee's ready execution of these various political commissions, its relationship with the OCB Working Group was never easy. It was apparently the Working Group which determined that the NSF should always be a co-signatory of satellite announcements with the Academy, not just occasionally and at the whim of the Academy, as Kaplan had wanted.[83] The Academy was further vexed at the snubs which seemed to be aimed at its still unsuccessful Vanguard project after the Army's success with Explorer 1. The President's congratulations were cabled, not to the Academy, but to the Foundation.[84] And the Academy's objections to hosting the Army's triumphant press conference in its prestigious building in Washington were overruled by the Working Group.[85]

The IGY Committee's most determined revolt over the functioning of the Working Group came in December 1957 in the form of:

> . . . a formal protest with the Director of the National Science Foundation with respect to the conduct of information activities by the Government not only on the recent unsuccessful launching but the entire test series. It was noted that this protest would be documented by newspaper accounts on VANGUARD. The U.S. National Committee representative particularly stressed the fact that the Deputy Director of the VANGUARD project, J. Paul Walsh, had, in his view, violated the OCB approved Guide Lines by his statement that the launching of the rocket was incidental to the test and that the purpose of the test was to place the satellite in orbit.[86]

In fact the ill-fated Vanguard test-sphere launch-attempt of 6 December led to abandonment of the previous custom of witholding most hard information about a test until after it had been concluded. Because it was impossible to prevent reporters at Cape Canaveral from listening in to the launch facility's activities on short-wave radios, the Working Group's Defense chairman (at this point Herschel Schooley) sanctioned an arrangement whereby:

> . . . the press was kept informed of the countdown progress and delays . . . in return for an agreement to withold their stories until the firing had been completed or definitely postponed. This procedure resulted in a bare minimum of speculative press accounts . . . [87]

The IGY Committee continued to criticize the arrangement as 'inconsistent with the agreed view of the Working Group', and eventually persuaded the Group to urge the Defense Department to restore the practice followed in early tests 'before the 6–incher was placed in the vehicle', under which 'no official announcement [was] to be made of the test date or launching schedule until after the test had been concluded'. But the Pentagon was very much a law, or a foreign policy, unto itself, and the Group could only ask its chairman 'to express this view to Mr. Snyder, Assistant Secretary of Defense for Public Affairs'.[88]

After that point the Academy was represented at only one further meeting of the Working Group, on 13 May 1958.[89] Its function in the decision-making process was being taken over by two new government bodies, the White House office of the newly-appointed Presidential Science Adviser, James Killian, and the National Aeronautical and Space Administration into which it was proposed to transform the existing National Advisory Committee for Aeronautics. Overall responsibility for the Vanguard satellite project and most of the project staff were transferred to NASA from the Academy and Naval Research Laboratory when the new agency formally opened for business on 1 October 1958. The Academy's departure from the Working Group in the spring of 1958 doubtless anticipated that foreseeable event, besides reflecting the scientists' growing disenchantment with the political role which they had earlier struggled to acquire. (The Working Group went on to enjoy its longest and most effective period of operation, under yet another title – 'On Outer Space', for the remainder of the Eisenhower administration.)

Plainly, Kaplan's empire-building demand that the Academy should be represented 'in groups concerned with Federal policy in this [satellite] as well as other areas' was never more than partially fulfilled.[90] But in addition to a presence at the OCB Working Group the Academy did achieve one other notable incursion into the process of government policy-making on space. Its president Detlev Bronk took part in two meetings of the National Security Council which discussed the satellite project, one on 10 May 1957 and the other on 10 October 1957, after the launching of Sputnik 1.[91]

\*                    \*                    \*

The Academy took little part in the work of a Special OCB Committee which began discussing 'any ideas . . . on regaining the psychological initiative in the cold war' from January 1958, with occasional participation by the NSF.[92] Proposals considered ranged from establishing the Peace Corps, through 'Union of Iraq, Jordan and Saudi Arabia', to 'Cultural Center in Angkor Wat'.[93] But emphasis was understandably placed on would-be demonstrations of technological prowess, such as excavating a harbour by means of nuclear explosives, a worldwide desalination programme, or beating the Russians to the Moon with an unmanned space probe.[94] The latter was strenuously advocated from mid-November 1957 by William Pickering, the director of the Jet Propulsion Laboratory who also chaired the IGY Committee Satellite Panel's Working Group on Tracking and Computation.[95] Dearborn liked it from the first, and suggested that 'it could be disguised as part of the satellite project'.[96] By February 1958 it was no longer being mentioned in Special Committee documents, probably because it had already been taken over, as the Pioneer programme of lunar probes using launchers based on the Air Force Thor and the Army Jupiter IRBMs, by the newly created Advanced Research Projects Agency in the Department of Defense.[97] Between August 1958 and March 1959 a series of disappointing shots:

> . . . failed to reassert American superiority in technical affairs. Instead, the Soviet Luna 1 . . . became the first unmanned craft to fly close by the moon. The distinction of first photographing the far side of the moon was also claimed by another Soviet machine, Luna 3, a short time later.[98]

What had started out as a Pentagon space programme was also handed over to NASA in October 1958, though execution of these space shots remained largely in the hands of the respective services and no attempt was made in the United States to portray them as part of the IGY. However lunar probes required three-dimensional tracking by powerful and widely separated radio telescopes. By April 1958 the Americans had notified the Royal Society, in terms that the late Frederick Dearborn would surely have approved, 'of a change in the experiments to be performed and the orbit to be followed in a new series of *artificial satellites* . . . [which] were very much dependent on the [newly completed] Jodrell Bank radio telescope'.[99] The fact that this was at first a Defense Department project was played down, and Jodrell Bank duly assisted in tracking all the Pioneer probes. However, the Royal Society declined the reciprocal offer of payload space on a future American satellite, because it would 'lead to a lowering of United Kingdom prestige'.[100] In fact the offer had been yet another political move originated by the OCB Working Group in an effort to minimize the harm done to American prestige by the early sputniks.[101]

Although not much involved with any of this, the US Academy did contribute substantially and over several years to a would-be spectacular research project which had strong political overtones as an intended demonstration of American technological capabilities, though it is not on any of Dearborn's lists of 'projects to regain the initiative' that has so far been declassified. This was the 'Mohole' project to drill several miles through the crust of the Earth to reach the Mohorovičić Discontinuity at the border between the crust and the mantle. While this was first mooted in scientific circles several months before Sputnik 1, the generous funding which it began to receive from early 1958 owed much to the intensified competition with the Soviet Union for scientific and technological prestige. In December 1957 Waterman reported to Elmer

Staats at the OCB that the 'Deep Hole' idea was a far better candidate
for 'the purposes of this exercise' than either solar energy or desalination
technologies, and enclosed a CIA assessment to the effect that, 'should it
become desirable on a crash basis, the US could probably drill through
the crust of the earth before the Soviets could'.[102] From April 1958 the
Academy accepted funding from the NSF to act as the project's offical
sponsor, and can hardly have been unaware of its cold-war aspects. This
suggests that any tendency for the Academy to withdraw from the more
political aspects of the space programme during 1958 was not part of a
general policy of replacing subservience to United States national interests
with a renewed professional independence.


## 7   THE POLITICAL PRICE

No matter how much Kaplan emphasized the independent and purely scientific character of the American IGY programme in general and the satellite
project in particular, the fact was that, in return for generous funding and
the provision by the armed services of development facilities for launch
vehicles, the scientists were expected to offer loyalty to United States
national interests and to relevant parts of the administration's national
security policy, particularly as we have seen its cold-war arrangements
for information-management. The scientists, or at least their leaders within
the Academy's IGY structure, did not merely put up with this. They began
by promoting the IGY to government in just such terms in 1953 and 1954.
When they were barred from the section of the administration's cold-war
apparatus which was tasked with exploiting the satellite project in pursuit of
national interests, they fought a hard bureaucratic campaign to be admitted.
When this succeeded, they played a full part in devising and implementing
aspects of the satellite project which had nothing to do with the objective
and impersonal goals of science. This political commitment was so fundamental to their thinking that even when arguing for independence Kaplan
put his case in terms of national interest:

> The freedom of science from governmental control is clearly involved
> in this matter, and the strength of our Nation's position internationally
> calls for clear indications that the Satellite effort is civilian.[103]

The source of this and similar confusions lay in the extra-scientific goals
that the Academy scientists had set for themselves with respect to the

satellite project. The first was that of securing a substantial say in the formulation and implementation of non-scientific national policy decisions affecting their own work. Whether or not they sincerely believed that by doing so they could safeguard their independence from government, the reverse was always the more likely outcome. In the circumstances of the day, talk about preserving the independence of science and about the scientists' professional responsibilities towards an international research programme (the terms of which they had had a large hand in drawing up) became little more than bureaucratic ammunition in a struggle for the policy-making upper hand between government agencies proper and a National Academy of Sciences which operated for a time at least as a quasi-governmental agency alongside the others.

The Academy's second extra-scientific goal was that of ensuring that genuine scientific achievement should become, as never before, a major dimension in the domestic pride and foreign prestige of the United States. As the satellite project came to dominate public perceptions of the IGY, however, it grew increasingly clear that both pride and prestige would be centred on easily comprehensible *technological* feats rather than on *scientific* progress as such. Since the majority of American and foreign citizens, then as now, had not received any higher education or in large parts of the world much education of any sort, it should have been obvious from the outset that spectacular demonstrations of technological capability would reach more people more effectively than would scientific feats like the discovery of the Van Allen radiation belts. Indeed NSC 5520, the official charter of the American satellite project, had clearly favoured a modest technological demonstration in the short term, to be followed at leisure by substantial scientific achievements. That document may never have been seen by the leaders of the US IGY Committee until after Sputnik 1, but even if it had been it would probably not have disturbed their élitist agenda. Four years after the IGY Killian preferred criticizing the public for its 'confusion between Soviet space *technology* and science' to criticizing his colleagues for failing to anticipate that confusion.[104] But Odishaw at least acknowledged that:

> . . . national interests did enter the space picture, even during the IGY. In the space program, certain aspects of national interests focused on the gross achievements of space systems – i.e. the capability of rocket engines. Science itself became a peripheral consideration despite the language used in statements originating in both the Soviet Union and the United States. Perhaps government interests were too great to avoid capitalizing upon space feats . . . [105]

It was doubtless even harder for American IGY scientists to recognize that they themselves had frequently chosen to play to the unsophisticated gallery, and not only in respect of the satellite project. The Antarctic research programme, for example, was not merely given the title of 'Operation Deep Freeze' by the US Navy but was sometimes also presented, even by Odishaw, as if its overriding goal had been to find out whether Russian or American scientists could record the lowest temperature at one of their respective stations.[106] (Here too the Russians 'won'.)

*          *          *

In the satellite case, the illusion that an American 'national science' might be allowed to run its own foreign policy and propaganda had depended upon two assumptions, neither of which turned out to be correct. The first was that Project Vanguard would be an unqualified success in both scientific and popular terms; the second was that the administration would be able to combine its sponsorship of prestigious American scientific endeavours with a strict policy of non-involvement with the details, however sensitive. Whether or not the IGY Committee had good reason to make those assumptions before October 1957, the sputniks brusquely overturned them both.

Even before that point, however, a shrewd observer might have predicted on the basis of Waterman's long experience in government that the National Science Foundation was unlikely to be defeated by the National Academy in a territorial struggle over such a salient project as the satellite. In his reply to Kaplan's letter of 24 July 1956 Waterman touched firmly on the political reality, which was that if the needs of 'policy' and 'science' in the satellite project were ever at variance the *former* would prevail:

> It has been and will continue to be the position of the Foundation that, *insofar as possible*, the earth satellite undertaking should be carried out as a scientific program, under the auspices of scientific organizations. This teamwork . . . should also recognize that the Foundation . . . is the Federal agency designated to coordinate *administrative and policy aspects* of the program in which other Federal agencies are involved. . . . Governmental authority . . . should only be involved in cases where this is necessary to secure Government approval of a particular program as a matter of national policy. In such cases the U.S. National Committee has already made it clear that *it would be guided by such policy in the national interest*. . . . This cooperation and good will must exist at all levels in the participating organizations

if our combined efforts are to be meaningful and productive *for science and for the Nation.*[107]

This was a considerable understatement, for 'governmental authority' and guidance had been involved in the American IGY effort from the outset, and the administration had been directly represented on the Academy's Committee to that end. In the case of the satellite project the administration had not only had the final say as to whether or not it should go ahead, but also on several details of its implementation and presentation.

In 1958 scientists anxious to serve both 'science and . . . the Nation' were presented with ample new scope for such activities, either in NASA or with the President's Science Advisory Committee. As for the ideal of the neutrality and independence of science, it was perhaps slightly better served by the departure of the IGY Committee from the OCB Working Group than it had been by its earlier struggle to join.

# 8 HISTORICAL ACCOUNTS

Historians and other writers describing the early Eisenhower space policy have often equated the administration's *declaratory* policy, which emphasized the scientific character of the IGY satellite before the sputniks and developed the point about its separation from missile projects after the sputniks, with its *action* policy, which as we have seen was more comprehensive and more subtle. Sometimes public statements made after 1957 are cited as evidence for the administration's real purposes before that date. But one of the first authors to do this did at least concede that Eisenhower might have had good reason not to damage public morale still further after Sputnik 1 by public lamentations of the sort to which his critics sometimes resorted. That consideration did not apparently rule out the conclusion that:

> Under the prevailing view [before Sputnik 1], national prestige was not at stake, and no one thought of mentioning national pride, i.e. the importance to national self-esteem that the country should be first in space.[108]

Others have used post-sputnik evidence to support claims about pre-sputnik policy without even hinting that the inference might not be sound. But the real problem with describing the Eisenhower administration's space policy as having been, from 1954 onwards, 'that space activities should be conducted solely for peaceful purposes', is not that that is false in itself.[109] Its misleading character consists rather in the implication that the policy was somehow philanthropic and internationalist. In reality American space policy before the sputniks was *both* directed towards 'peaceful purposes' *and* intended to combine the cold-war benefits of a prestigious American scientific achievement within the IGY with longer-term military benefits via satellite reconnaissance. As far as the administration was concerned the prestige goals of its public satellite project and the indirect military dividend simply *were* 'peaceful purposes', and ones for which it was entitled to employ either total secrecy or a certain delicacy of language as it saw fit.

Another frequent assumption has been that if the administration (supposedly) intended the IGY satellite project to be 'conducted for scientific purposes only' that was somehow evidence of its 'low estimate of the political significance of a satellite launch'.[110] The problem with this is that, unless it is a mere tautology, it implies the contradiction that the Eisenhower administration was both indifferent to national prestige and determined to uphold it on a basis of genuine scientific achievement. Since Van Dyke, the first author cited in this section, carefully explained as long ago as 1965 that the latter was in fact the Eisenhower policy on scientific prestige, later scholars had little excuse for confusing the issue.

Furthermore, the guidance extended by the OCB on public-relations aspects of the IGY satellite project, with regular help from the IGY Committee, shows that the Eisenhower administration was by no means content to sit back and wait for the normal scientific prestige to be gained from launching what was expected to be the world's first artificial satellite. Indeed the propagandistic or cold-war aspect of the satellite project received at least as much input from senior levels of government as any other, and possibly more. But though most of its minutes were declassified as long ago as 1976 the pre-sputnik activities of the OCB Working Group have yet to be acknowledged by space historians.[111]

$$*\qquad\qquad*\qquad\qquad*$$

If the evidence for the prestige element in the early space policy of the Eisenhower administration has been slow to emerge, that can hardly be said for its military element. By late 1954 or early 1955 the President had decided to support the eventual development of military reconnaissance satellites as the natural successors to the U-2 spy plane, which had its first test flights in the second half of 1955 and began operating over the Soviet Union a year later. Commentators began to suggest the existence of a covert reconnaissance satellite project immediately after the July 1955 announcement of the IGY project, by linking the latter with the President's recent arms control initiative for 'Open Skies'. Official denials were issued, but two years later, after Sputnik 1, *Aviation Week* disclosed the existence of the 'Pied Piper' reconnaissance satellite project – the full development stage resulting from Project Feed Back.[112] And in January 1958 General Schriever confirmed to the Preparedness Investigating Subcommittee that 'there was a lot of interest at different sources in the government for an advanced reconnaissance system . . . Now since Sputniks, there has been, of course, a desire to accelerate this program.'[113]

The early evidence that the Eisenhower administration had a military space policy before Sputnik 1 has been supplemented by several excellent studies, but few have discussed its interaction with the IGY satellite project.[114] McDougall was probably the first to point out, as late as 1985, that NSC 5520 not only required the IGY project not to interfere with the national missile programme but also not to hamper work on future military satellites which was already in hand.[115]

&ast;               &ast;               &ast;

The claim that the Eisenhower administration failed to appreciate the prestige aspect of Project Vanguard has been linked with the non-interference requirement in NSC 5520 to support the view that administration policy was directly responsible for the Stewart Committee's selection of the Vanguard launch system, and therefore for the American humiliation in October 1957. But this argument has relied on confusion and inattention to the evidence. The confusion consists in assimilating the non-interference requirement, which was clearly part of the administration's action policy for Project Vanguard, to the declaratory policy of emphasizing the civilian and scientific aspects of the project in public statements. By conflating these two distinct elements of policy some scholars have continued to obscure the fact that the administration intended to achieve certain non-scientific and indirectly military goals through Vanguard, though not at the expense of its top-priority missile programmes, and have preserved the long-established fiction that the administration tried to make Vanguard *in fact*, and not just in appearance, 'a purely scientific endeavor'.[116]

The considerations which influenced the Stewart Committee's decision were placed on the record in the Congressional report on Project Vanguard published in 1959.[117] Its findings were comprehensively reviewed by Constance Green and Milton Lomask in the official history of the project in 1971. But Green and Lomask appear to have added their own crucial but highly tendentious conjecture that the Stewart Committee was also 'Instructed . . . to regard the satellite program as purely scientific *rather than* politically significant'.[118] Once again the possibility that it was seen as *both* was overlooked.

The picture of the Stewart Committee as having had no option but to endorse the allegedly 'civilian' Vanguard in order to satisfy the administration's quixotic altruism about space is still very much alive.[119] But its survival owes much to the intensity of the partisan criticism which burst out against the administration in the wake of the first sputniks, and which interacted strongly with the earliest and most influential Congressional investigation of the sputniks affair, the subject of Chapter 12.

# Part V  Judgements

# 12 Post-Mortem

## 1 THE POLITICAL BACKGROUND

On 25 November 1957 the Preparedness Investigating Subcommittee began public hearings for an 'Inquiry into Satellite and Missile Programs', for which its chairman, the Democratic Majority Leader Lyndon Johnson, had initiated the necessary preparations immediately after Sputnik 1. The Subcommittee hearings covered a great deal of ground and contain much raw material of value to the space historian. However it will be shown in this chapter that their reputation as a comprehensive account of the facts behind the US-Soviet balance in respect of long-range missiles and unmanned satellites, as of late 1957, has been considerably exaggerated.

The tendency to exalt the achievements of the Subcommittee beyond their actual merits began with the senators themselves. Senator Saltonstall opined that 'of all the committees that I have sat on since I have been here, I know of none that have gone into questions more thoroughly than has this committee'.[1] Senator Johnson called the inquiry 'one of the most thorough in my memory'.[2] He sought to contrast its findings, based on 'sworn testimony by top scientists, leading industrialists, and Government and military officials',[3] with the course of action taken by the administration before the sputniks. In his view the latter had been determined by 'national policy' alone, by which he meant the arbitrary intervention of inexpert civilians in high office, so that 'the evaluation of the importance of control of outer space made by us [sc. the US government] has not been based primarily on the judgment of men most qualified to make such an appraisal'.[4]

\*　　　　　\*　　　　　\*

But Johnson's handling of the Subcommittee cannot be understood apart from its domestic political context. By 1956, if not earlier, the Democrats were hungry for a truly national issue which they could make their own. In April 1956, for example, Johnson was advised by his aide Gerald Siegel that:

> If we move forward vigorously on an affirmative program, the second session can be very productive and the people can be made to recognize it as such. There are no political issues of a national character and we cannot make any by wishing them.[5]

Likewise in March 1957 another of Johnson's advisers, George Reedy, warned him not to try to fabricate a national issue:

> The reality is that issues must be *recognized* and *exploited*. But they cannot be spun out of the blue . . . Granted that no 'great issue' has emerged thus far. The point still remains that we will not create one as a sculptor creates a statue.[6]

But in fact Johnson's advisers had more than once suggested *before* the sputniks that the Preparedness Subcommittee would do well to look into the state of the American missiles programme. The most recent such proposal was made in February 1957, but Johnson ignored it until after Sputnik 1.[7] The peak of his interest before that date appears to have been aroused by the possibility that the removal of the von Braun team and other portions of the missile programme from Fort Bliss to locations outside Texas would cause local unemployment in El Paso.[8] In line with the evidence reviewed in Chapter 2, Siegel later stated that 'neither Mr Johnson nor any other Senator, to my knowledge, had given any really serious thought to either missiles or to anything like a civilian space potential . . . prior to the launching of Sputnik'.[9]

*          *          *

The national issue which had Johnson and the Democrats in its toils in 1956 and 1957 was that of racial integration in the southern United States, a social development strongly opposed by the white conservatives who formed the party's traditional power-base in the region. Johnson's advisers urged him to avoid any hint of a pro-segregation stance which would deny him his party's presidential nomination for the 1960 election, and to pose instead as a national conciliator, supporting civil rights legislation in the Senate while modifying it sufficiently to appease the sharper resentments of the South.[10] As Senate Majority Leader Johnson in fact played a leading role in securing passage of the 1957 Civil Rights Act. But the segregationism entrenched within his party was skilfully exploited by Republicans, making it impossible for a prominent Southern Democrat to establish himself as an alternative national leader to the still popular incumbent President. Indeed, on the day Sputnik 1 was launched political attention in the United States was focused on the town of Little Rock, Arkansas, where the National Guard had been mobilized to enforce a Supreme Court ruling in favour of racial integration in the public schools. Eisenhower had even been obliged to cancel plans for high-profile visits to the missile development facilities at Cape Canaveral and Huntsville that summer, because both were in the South, and the situation was deemed so volatile that the President could not afford to associate himself publicly with any military units in the region.[11]

As Reedy described the Democrats' problem soon after Sputnik 1:

... in the integration issue [the Republicans] have a potent weapon which chews the Democratic Party to pieces so efficiently that it cannot be an effective opposition. ... The only possibility is to find another issue which is even more potent. Otherwise the Democratic future is bleak.[12]

Not surprisingly, Reedy was quick to seize on the Russian satellite as a possible solution, and one ideally suited to advance his man's career:

*If the issue has merit, the politics will take care of themselves* ... A central figure must be a Senator who has a reputation for statesmanlike, non-partisan investigations of defense and who is not involved too heavily in the emotional issues surrounding segregation ... At least in the early stages, implications should be scrupulously avoided ... Leave the implications to the American people who should be regarded as a jury ... There is, of course, a partisan advantage to turning the attention of the American people to such an issue.[13]

While Johnson had almost certainly made up his mind 'to plunge heavily on this one', in Reedy's phrase, *before* receiving this advice, the memorandum attests the strong political motivation that he had for doing so. Within a week, some of the more publishable material from it was being widely reported, after it had featured in a speech written by Reedy that was delivered by Johnson in Austin.[14]

## 2 SUBCOMMITTEE PROCEDURES

The published hearings, together with papers in the Johnson Library and elsewhere, give some indication of the methods that were followed in preparing and presenting the Subcommittee's investigation for public view. Reedy originally recommended that the Subcommittee confine itself, at first, to 'gathering the facts and presenting them to the public'.[15] This should be done methodically, with due attention to scientific and historical detail:

... some expert must in plain language describe a satellite. He must also testify as to the scientific and military significance of these satellites now traveling in the air. [*sic*] Someone must testify how the Russians were able to accomplish this feat ... Where do we stand with respect to missiles? ... Why are we so behind?[16]

But Gerald Siegel argued for a very different approach, to be built around a recent article by Edward Teller which:

> ... spells out ... the appropriate new roles and missions which our military service should prepare for, ... establishes the desirability and necessity of preparation for both limited and all-out nuclear war ... [and] describes the kind of preparation our civilian economy must make if it is to survive: (a) a nuclear attack and (b) the period of devastation in the country following such an attack.
>
> It is my judgment that it would be desirable to have the Preparedness Investigating Subcommittee hold hearings on these broad policy questions. They are of greater importance even than the missile-satellite programs, which become only a small part of these broader questions.[17]

Siegel suggested that the hearings might proceed by inviting Teller to recapitulate his proposals, and then producing prominent political figures or academic experts, such as former Secretary for Air Thomas Finletter, former Secretary of State Dean Acheson, Dulles (unidentified), the new Secretary of Defense Neil McElroy and the Harvard strategic studies scholar Henry Kissinger, to comment and enlarge upon them.

The pattern of the opening days followed Siegel's rather than Reedy's model. Reedy now attributes that development to a seven-hour briefing at the Pentagon which'he attended in support of Senators Russell, Johnson and Bridges on 2 November 1957 (the day before Sputnik 2). The session had featured 'very depressing' numerical comparisons of Soviet and American strategic forces, which 'had a tendency immediately to get us into issues of rocketry and strategic missiles, and to avoid the broader issues of space'.[18]

According to Reedy the Subcommittee's chief counsel, Edwin Weisl, chose Teller for the key role of opening witness because:

> He had a colorful and dramatic story to tell – O God it was colorful and dramatic. I didn't like it myself. .... I thought it was not a logical presentation of what we were after. ....

Weisl saw it entirely in terms of a dramatic presentation. Once a thing ceased to be dramatic, I think he sort of lost interest in it.[19]

The basis for the Subcommittee's selection of other witnesses is hard to reconstruct, but some at least were chosen as much for political effect, on the basis of articles or interviews published before they came to the hearings, as for any expertise they possessed on satellites or missiles. Reedy's comment about the emphasis that was placed on dramatic presentation is substantially confirmed in the record by the choice and handling of witnesses such as Admiral Hyman Rickover, Alexander Seversky and John Gleason, Commander of the electorally powerful American Legion.

The Subcommittee staff also polled selected industrialists, educators, scientists and officials with a questionnaire asking what had been the main problems in missile and satellite programmes and what should be done to remedy them. While this was obviously a legitimate and effective procedure, no thematic analysis was ever published of the content of all replies, both negative and positive, and selected passages could be read out in the public sessions as much for their dramatic as for their informative content. The astronomer Fritz Zwicky, for example, had little to offer on satellites or missiles but seized the opportunity to boost his own experiments on rocket-launched 'artificial meteors' as work 'which will pave the march of free men into space ahead of the Bolsheviks', and to denounce his less enthusiastic scientific colleagues as 'fellow travelers, slackers, and actual pinkoes'.[20]

It is impossible to prove the hypothesis, but the non-appearance of Clifford Furnas, who had played a major role in the satellite project, first as a science adviser and then as the senior science official at the Pentagon, may have been a result of the article which he wrote for *Life* after Sputnik 1. For although Furnas had joined the general chorus of criticisms directed against the administration's handling of the satellite project, he had *also* suggested that attitudes and decisions under the previous, Democratic, administration had contributed to the American failure.[21] To judge from the balance of evidence presented, that was a point which those who were managing the hearings were most unwilling to have aired on their particular platform.

\*　　　　　\*　　　　　\*

The political sleight-of-hand to which the Subcommittee sometimes resorted was vividly described in a contemporary diary entry by Oliver Gale, a senior official in the Office of the Secretary of Defense and therefore an admittedly partisan witness:

February 26 [1958] – The Johnson Committee has hit upon a very successful formula. They have set a series of hearings, a month apart, at which the Secretary and his people will report on the advancing state of our preparedness. They ask Defense to submit a week or so in advance a written report on what we are doing to catch up with the Soviets . . . This – a highly classified document – is delivered by me to my opposite number, Jerry Seigel [*sic*], of the Committee staff. Then, following the hearing, they issue to the press a report in which they urge Defense to do the very things we have said we were doing. The picture is clear: they are directing Defense, leading the nation in its frantic rush to reduce the state of peril, and we are gratefully – or perhaps even reluctantly – doing as we are told.

We go along partly because we have no choice, and partly because these are the same individuals who have to approve our military budget and we can do nothing but lose if we fight them.

Meanwhile critics of the Administration have been having a field day. The team headed by [counsels] Weisl and Vance probe, in their pre-hearing interviews, for disaffections and pet projects of the various military personages. These are bound to be many: no one is getting all the funding he wants, each thinks his specialty is the most important (if he didn't he wouldn't be able to do his job properly); each thinks that if there is a job to do, such as airlift, his service should be the one to do it, etc. These views, many of them in conflict with the announced positions of the Secretary, are brought out in the hearings, and they make headlines.

Also, the contractors get prominently into the act. Here is a chance to tell Congress that one's own weapon system is just tops, and should have a lot more money spent on it. . . .

But it's all part of the process, and no one is seriously hurt.[22]

Whether or not Gale's closing remark was accurate, it can certainly be wondered whether the truth was particularly well served by such manœuvres, to say nothing of the contemporary effectiveness and later reputation of the Eisenhower government.

## 3  OMISSIONS

Although Weisl and his staff presented the senators with some evidence of historical and contemporary fact, they did so in a random and incomplete fashion, and never produced an objective description of the satellite project and its organizational history of the sort proposed by Reedy. Even in his opening remarks Johnson made it clear that what he wanted was simply confirmation of what he chose to regard as common knowledge, in the form of:

> . . . a clear definition of the present threat to our security, perhaps the greatest that our country has ever known. . . .
> It is not necessary to hold these hearings to determine that we have lost an important battle in technology. That has been demonstrated by the satellites that are whistling above our heads . . . [23]

Johnson dinned it into witness after witness that he had been called primarily to help redeem the situation, and only secondarily to analyse its origins and nature. It was a device which left little scope for anyone to query the alleged severity of the crisis:

> But, without getting into the dollars now or the budget now, the physical situation now, the chairman of this committee is going to insist that each witness say, in as plain, simple terms as possible, where he who listens can hear, the answer to one simple question: What can be done? [24]

The Subcommittee naturally accepted that Defense Secretary Neil Mc-Elroy's experience in senior management at Procter & Gamble had not equipped him to inform them about 'all the things that have happened in the past'. [25] But the senators also declined to consult the memories of better-qualified witnesses. When von Braun touched on the sensitive issue of the missile programmes of the Truman administration, the matter was not pursued, and the obvious witness, Truman's missile manager Kaufman Keller, was not even called. [26] The question 'If [von Braun] was stopped, who stopped him?' was not put to the veteran missile designer himself, nor to anyone with first-hand knowledge of the period from 1945 to 1952, such as Vannevar Bush, but only to Deputy Secretary of Defense Donald Quarles, whose answer made it clear that he was scarcely qualified to give one. [27] In a similar vein, the important question of what had become of the satellite programme announced in Secretary of Defense Forrestal's 1948 report was never asked.

Even the facts about the satellite decisions of the Eisenhower administration were scanted. Nelson Rockefeller was asked nothing about the political or technical origins of the American IGY satellite project, with which he had been closely concerned as a White House adviser in 1955, but only about the recommendations on security matters of a blue-ribbon panel on government reorganization which he had recently chaired.[28] Knowledgeable members of the Stewart Committee or the US IGY Committee, or both, such as Richard Porter, Joseph Kaplan and Clifford Furnas, who had taken part in the original selection of Vanguard and in forming the crucial recommendation not to approve a parallel Army project in 1956, were not called. Former Secretary of Defense Charles Wilson, who had retired just after Sputnik 1, and who would certainly have conducted a caustic defence of the administration's policy of giving priority to missiles, did not appear. Neither did the former Assistant Secretary for Air Trevor Gardner, who had opposed both the satellite project and the Army's continued interest in long-range missiles (which *was* well represented both on and off stage at the hearings), and who might have had something to say about the state of the American missile development programme as it was inherited from the previous administration in January 1953.

The Subcommittee did learn something, though not very much, about the arguments in favour of a satellite programme which had been presented by Killian's Technological Capabilities Panel in 1955. But none of the senators sought to clarify the relationship between early research satellites and later satellites for military applications, which the TCP Report had certainly discussed, even when they were invited to do so by McElroy with the remark that 'the type of satellite that the military was interested in required development that went way beyond the kind of instrumentation that is considered right for the IGY satellite'.[29] Still less were they interested in such plausible explanations for the satellite débâcle as Eisenhower's favourite, the time-consuming preference of the scientists for excessive levels and quantities of satellite instrumentation, which was expressed to them by Quarles.[30]

## 4   THEMES AND FINDINGS

In keeping with its haphazard approach to historical fact, the Subcommittee never produced a report on the causes of the American failure to launch the world's first satellite, though many were of course suggested by witnesses.[31] Its own eventual 'Statement' merely listed seventeen loosely-worded recommendations, in no particular order, and without paying any attention to the fact that, while the TCP Report had certainly recommended a satellite project of some kind in 1954 at about the same time that other elements within the administration were reaching the same decision, it had also firmly endorsed the policy of the day. Typically, it was possible for this to be stated far more clearly at *another* Congressional hearings by the Assistant Secretary of the Navy for Air, Garrison Norton:

> The ground rules specified that this program [Vanguard] was to have a priority below that of the ballistic missile and was not to interfere in any way with the ballistic-missile effort. That decision was based on the belief that we were not ahead of the Russians in missile development. Hence, we could not allow any interference with our own programs, a decision with which I am certainly in accord.[32]

Since this policy was widely held to have been a major factor in the decisions affecting Project Vanguard, the Subcommittee might have been expected both to have established the origins and reasons for the policy, and to have made their own positions clear with respect to it. They did neither. In a similar vein, when they recommended:

> 8. Step up production schedules of Atlas, Thor, Jupiter, and accelerate the development of Titan . . . ,

they neither acknowledged that such production would not have been in question by 1958 had it not been for the intensive development efforts of recent years, nor stated whether it was more or less vital than, for example:

> 5. Modernize and strengthen ground and naval forces . . .
> 7. Pour more effort into our antisubmarine program . . . [or]
> 16. Put more effort into the development of manned missiles.[33]

The failure to analyse thoroughly the administration's policy of giving priority to missile programmes, at the possible expense of the satellite, was the more remarkable because, when it suited him, Johnson went to some lengths to claim that administration policy had been the prime cause of the national humiliation, as in the following intervention:

> *General Schriever*: The matter of the satellite and whether or not . . . we could have launched one earlier, I don't think is so much a matter of the organization down there as it is a matter of attitude on the part of the Government.
> *Senator Bush*: I did not mean to open that question. I agree with that.
> *Senator Johnson*: If the Senator will permit an interruption there, what you are saying is, it is a question of policy, isn't it?

*Schriever*: That is correct . . .
*Johnson:* The President has stated that and others have repeated it.[34]

\*          \*          \*

But this exchange and others like it in the hearings can only be properly understood once it is realized that Johnson was not primarily using 'policy' and related terms to refer to detailed aspects of the administration's past course of action, such as its emphasis on missile development. This was not a matter of non-partisan forbearance, although that could usefully be simulated in passing. It stemmed, rather, from Johnson's *a priori* conviction that the sputniks crisis had sprung from a single, comprehensive cause, the failure of Eisenhower and his senior advisers to achieve a true appreciation of 'the significance of the control of outer space'.[35] The purport of such expressions was a fusion of two related ideas, one derived from von Braun, the other from Teller.

Since 1952, at least, von Braun had been pushing a military doctrine of space which was shortly to be designated 'the Panama Hypothesis'.[36] The idea was that somewhere in space, either on the Moon or as a missile-carrying space station or in some other form, there existed, potentially, the definitive solution to the security problems of the United States and its allies. Whatever form this solution might take, and wherever it would turn out to be located, there was little chance of laying hold of it if the Soviet Union was allowed to go everywhere and do everything in space first. Johnson's belief in this vague but powerful idea was manifested in language such as the following:

> If, out in space, there is the ultimate position – from which *total control of the earth* may be exercised – then our national goal and the goal of all free men must be to win and hold that position. . . .
> From space, *the masters of infinity* would have the power to control the earth's weather, to cause drought and flood, *to change the tides* and raise the levels of the sea, to divert the Gulf stream and *change temperate climates to frigid*.[37]

But the Panama Hypothesis about space was merely the astronautical version of a more general idea repeatedly expressed by Edward Teller, which was that a potentially decisive role in international affairs, and above all in the security of the United States, both could and should be played by advanced military technology. At the Subcommittee hearings his specific enthusiasms were for nuclear-powered missile submarines and for a major development programme for anti-missile defences, both passive and active.[38] He declined to guess what would be the military gains from space exploration, a subject he knew little about, but merely avowed an absolute confidence that there would be some.

With time, Teller's would become a familiar and more readily resistible voice, repeatedly combining deprecation of any possible political solutions with insinuations of an imminent technological victory in the nuclear arms race by means of devices ranging from radiation-free H-bombs to nuclear-pumped X-ray laser weapons. In the climate of the sputniks crisis, however, Johnson either was or affected to be convinced that a military-technological grail of some kind was out there in space for the taking. This led him into such extravagant statements as 'the missiles which are so dreaded today may prove obsolete almost as soon as they become realities', a wildly optimistic surmise about ballistic missile defence.[39] His faith in the Panama Hypothesis also lay at the root of the masterly ambivalence with which he alternated between such hyperboles as 'I guess there is enough blame to go around for everyone in America,' and, outside Congress, 'How long, how long, O God, how long will it take us to catch up with Russia's two satellites?'[40] on the one hand, and admonitions to his colleagues that 'I don't think this is any time to panic or be hysterical,' on the other.[41]

## 5 ABSTAINING FROM HISTORY

In view of the regularity with which, at the time, people likened the sputniks crisis to Pearl Harbor, it is interesting to compare the treatments which the two episodes received in the Congress. The Pearl Harbor Inquiry was held several years after the event and, however imperfect, went into considerable detail about most aspects of the Japanese operation and its political and strategic background. It was exploited to some extent for partisan purposes by the Republican minority on the Joint Committee, whose party had long been waiting for an opportunity to take their political revenge on President Roosevelt, even posthumously, and thereby also to restore their own political credibility, much weakened by the long Democratic ascendancy. The Preparedness Subcommittee hearings after the sputniks were less overtly partisan and at the same time less interested in history. This actually freed the Democrats to echo and thereby to reinforce the popular assumption of the day, that most or all of the responsibility for the setback represented by the sputniks lay with the Eisenhower administration, without being called upon to prove it. It also left the field of space history wide open, to begin with, for amateur contributions from spaceflight enthusiasts, who were naturally shorter on hard information than on the conviction that their pet concern had been unnecessarily and irresponsibly slighted by the authorities. For their part, senior administration officials and advisers neither wanted nor were then in a position to rebut such attacks.

Even a quantitative comparison is revealing. The Pearl Harbor Committee heard about 45 witnesses on 68 days over a period of 98 days, and published their testimony in eleven volumes, totalling 14 765 pages, together with 4780 pages of exhibits and many scores of maps and plans. Numerous messages passed by American forces in the Pacific during the period in question, and indeed the effects of almost every round of ammunition fired in the Japanese attack, were examined. A further 18 volumes of related previous hearings and exhibits were attached.[42] The Preparedness Subcommittee heard from more witnesses (about 75) but far more briefly, on 30 days over a period of 110 days, with three one-day follow-up sessions to review the progress of its recommendations. The published record, in three parts, is 2476 pages long.[43] While this was no small output in terms of Congressional activity, it represented nothing like the extensive historical coverage achieved by the Pearl Harbor Committee.

*              *              *

A few days after Sputnik 1 Johnson telephoned Congressional researcher Eilene Galloway to tell her that he intended 'to make a record in outer space'.[44] The Preparedness Subcommittee hearings were extremely effective in that respect at least. Indeed, only when it is understood that the hearings were aimed primarily at achieving a transfer of the policy-making initiative does the Subcommittee's relative indifference to facts become intelligible. By January 1958, when Johnson had secured the chairmanship of the Senate's prestigious new space committee, even his minimal need for historical knowledge was at an end:

> The past is already for historians. Let us seek solutions so that the future may be written by free world historians.[45]

But Johnson was able to dominate the hearings with his favoured explanation for the sputniks crisis, not merely through his personal authority and great political skills, but also because the line he followed so completely met the interests of other parties, notably the scientists and the military, each of which groups had good reason to prefer a vague indictment of national leaders to a detailed examination of their own past actions. The only witnesses ever placed on the defensive were government officials, above all Donald Quarles as a former head of Defense R&D and the most senior really knowledgeable official still in post, and William Holaday as Director of Guided Missiles. The director of Project Vanguard, John Hagen, was treated with sympathy as a man struggling against unfair odds, and the new Secretary of Defense was also well received, if only as someone doing his best to repair the implied mistakes of his predecessor, Charles Wilson. As mentioned above, neither Wilson himself, nor other key officials who had resigned before the sputniks, such as Gardner and Furnas, were called. Nor, by constitutional tradition, were members of the White House staff, many of whom would have known where certain skeletons were buried in the record of policy formation. This procedure meant that almost all present could unite in deploring the shortcomings of recent government 'policy' or 'attitudes' without the embarrassment of determining exactly how such policy had been formed and exactly who had subscribed to it at the time. In his pursuit of bigger game Johnson was prepared to let the small fry go, provided they went along with his version of events.

*       *       *

Other Congressional hearings made a slightly better job of finding out the facts. The House Armed Services Committee held an 'Investigation of National Defense Missiles' in January and February 1958 for which its chief counsel, Robert Smart, took pains to establish such relevant matters as the membership of the Stewart Committee and an accurate if incomplete chronology of the decision-making process which had led to Vanguard, and at which witnesses were given the opportunity to point out the importance of the delay on ballistic missiles under Truman and the fact that the Eisenhower administration had had an Air Force reconnaissance satellite project in hand for some years. Hagen was able to emphasize that the Stewart Committee's original selection of the Navy's Vanguard proposal had basically been determined by the consideration that the Army's Orbiter alternative offered a mere 5 lb payload as compared with Vanguard's 20 to 30 lb. Assistant Secretary Norton could challenge the Committee to fault the administration's policy of giving top priority to its missile programme in terms that Johnson would hardly have tolerated (above). But much continued to be made of the mythical 'statement by a Russian scientist [Sedov in 1955] . . . announcing the intent of the Russians to try to launch a sputnik 6 months before the International Geophysical Year'. And Representative Richard Lankford earned nothing more daunting than supportive laughter from his colleagues as he led the Army witnesses into pinning the entire responsibility for having ignored that (non-existent) piece of intelligence onto Donald Quarles.[46]

A new House Select Committee on Astronautics and Space Exploration held its first hearings in April and May. Important and previously neglected witnesses such as Keller, Furnas, and leading members of the US IGY Committee were called, in Keller's case to give his only Congressional testimony after Sputnik 1. But Furnas was still not asked about his role on the Stewart Committee or subsequently as the Pentagon's head of R&D during Project Vanguard. These hearings established that Keller's priority under Truman had been for air defence, and that he had actually delayed development work on the propulsion side of the Atlas project pending progress with the new technologies of guidance and warhead re-entry. They also learned from Herbert York, chief scientist at the Pentagon's new Advanced Research Projects Agency, that the Soviet lead had effectively been established 'as soon as they started working on bigger engines than we were working on. That must have been in the late forties.'[47] And although the ABMA commander Major General John Medaris repeated the Army claim that it could have launched a satellite with a Redstone-based launcher as early as September 1956, von Braun made the point that they would have to have been very lucky to do so at that time. He also stated that the Vanguard team had exceeded his expectations by launching their first successful satellite in March 1958, a mere 30 months after starting work, a point which was underlined by several Navy witnesses.[48] Clearly the successive launches of the first American satellites – Explorer 1 on 31 January, Vanguard 1 on 17 March and Explorer 3 on 26 March – had done much to reduce anxiety and increase objectivity in Washington. But Committee members still pressed for assurances that the Russians must also have had 'plenty of failures' in their early space programme.[49] And the claim that the humiliation of the United States had flowed solely and directly from the administration's decision not to let the IGY satellite project interfere with missile programmes was still as popular as ever.[50]

*         *         *

In the historical literature the Preparedness Subcommittee hearings have clearly made a deeper impression than any others, perhaps because they came first and were thus contemporary with the height of the domestic crisis over the sputniks, or perhaps as a result of Weisl's constant attention to their dramatic potential. The negative consequences of this for the factual element in the historiographic tradition were suggested by Eilene Galloway, who had worked as a Congressional researcher for both the Preparedness Subcommittee and the House Committee on Astronautics, when she remarked that 'the Senate would never have had a thorough hearing like this [House, Astronautics] on outer space, itself, because you couldn't get them to sit still long enough in meetings because they were so busy'.[51]

The casual approach to factual enquiry that was adopted at the Subcommittee hearings is evident from a single, in itself quite minor, example. Senators naturally wanted to discuss the Advisory Group on Special Capabilities that had been chaired by Homer Stewart, but no one was able to remember Stewart's name until the third day of the hearings. Those who were groping for it were simply left to do so, in spite of the fact that Stewart himself was present throughout, having been appointed senior scientific adviser to the Subcommittee with instructions to 'Sit behind Senators at hearings, to suggest questions and interpret testimony.'[52] And even when a list of members of the Advisory Group was belatedly produced, in December, the name of Clifford Furnas was omitted. It bears repeating that neither he, nor Stewart, nor any other member of the Group was ever required to testify before the Subcommittee.

Such omissions have seldom been noticed by the historians, who have preferred to echo the Subcommittee's self-esteem by hailing its hearings as 'thorough', 'monumental'' 'a minor masterpiece', or 'a textbook example of what a Senate investigation ought to be'.[53] The NASA Historian Eugene Emme and at least one other even wrongly credited the Subcommittee with having recommended 'the establishment of an independent space agency'.[54] But in fact Johnson never really sought to 'elevate the hearings into the realm of space and away from interservice battles in the Pentagon', and only one of the Subcommittee's seventeen recommendations, for the development of a rocket engine with one million pounds thrust, was directly concerned with astronautics.[55]

However some historians have recognized the partisan political dimension of the Subcommittee's activities. Enid Schoettle explained in an early study that Johnson's campaign for a major effort in space was conducted 'for his own and his country's welfare', and that his avoidance of the 'arrant partisanship of both liberal Democrats and conservative Republicans' was directed at a more subtle political goal, that of keeping the administration 'implicitly, if not explicitly, on the defensive' while proclaiming the message 'that the country lacked leadership and that the Democrats would provide it, whether or not the administration went along'.[56] John Logsdon at least remarked that Johnson 'wanted to become President'.[57] And Walter McDougall points out that 'the whole story of the Vanguard/Jupiter decision did not come out', drawing a picture of 'witnesses [who] rose to confirm the committee's suspicions and provide quotes for the next day's front pages'.[58]

*        *        *

But undoubtedly the most important legacy of the Preparedness Sub-committee hearings to the historiographic tradition has been the habit of accepting without critical examination Johnson's original partisan claim that 'policy', as made by a handful of senior members of the Eisenhower administration, had been the unique historical cause of the initial selection, poor management and allegedly 'low priority' of Project Vanguard, and hence of the setback to American interests represented by the sputniks. No historians to date have noticed that, by concentrating his fire on the alleged failure of the administration to 'understand' the strategic value of space, Johnson successfully concealed the central contradiction in his own position, which was that he was simultaneously accusing the administration both of *failing* to give sufficient priority to missiles, and of giving them such an *absolute* priority as to cause the loss of American prestige through the poor performance of the satellite project.

The uncharitable explanation for the long consent of the space historians to the rather shallow view of the sputniks crisis that was taken by the Subcommittee is that they have too often shared the prejudices of a 'pro-space' lobby which had a strong interest in scapegoating the Eisenhower administration. A more charitable one might be that, whatever the Subcommittee hearings did or did not do by way of *investigating* history, they *made* more history than any other comparable political process in post-sputnik Washington, and this aspect of them has simply captured the attention and imagination of all subsequent American writers on space. Perhaps historians are only slightly less liable than other people to allow the terms in which they read the past to be set by the protagonists of history themselves.

# 13 Versions of Events

The historiographical tradition which has been established in respect of the sputniks crisis can be seen as an agglomeration of several fairly similar versions of events and a denial of any which differs from them. A truly alternative view has recently been presented, but as a rather minor theme in what is arguably the only work of genius so far written in the field. This concluding chapter reviews those versions of events, roughly in the order of their appearance.

## 1  TRUMAN'S VERSION

It was mentioned at the end of Chapter 6 that during the sputniks crisis the former US President Harry Truman declared that all would have been well with the United States in respect of long-range missiles and, by implication, satellites, if the Eisenhower administration had not mishandled the effective rocket development programme inherited from its predecessor. Little substance was found in this partisan view of events. But by neglecting the difficult issue of remote causality and concentrating instead on a period of three or four years before Sputnik 1, space historians have tended to give it a certain tacit credence.

Counter-factual historical hypotheses can never be proved beyond doubt. But those protagonists or historians who have asked themselves the question have generally concluded that if the United States had started a full-scale development effort for large ballistic missiles only slightly earlier than 1954 it would almost certainly have been the first country to place artificial satellites in orbit. This judgement implies that a large, if incalculable, share of the responsibility for the outcome of the first space race belongs to the Truman administration. It has been supported here by two subordinate findings. First, that in the late 1940s Truman had access to sound intelligence about the Soviet lead in rocketry. Second, that several shrewd and prescient technical studies and policy papers advising about the feasibility and the potential value of artificial satellites were prepared by governmental or quasi-governmental agencies under his administration. The tendency of space historians, however, has been to assign almost all the responsibility for the sputniks crisis to the policies of the Eisenhower administration, and very little to those of its predecessor.

## 2   AN EISENHOWER VERSION

Eisenhower published the second volume of his memoirs ten years after his administration had given the go-ahead for an American IGY satellite. One chapter dealt with the sputniks crisis.[1] To his own recollection, Eisenhower was first told about the feasibility and desirability of satellites in 1954. That would be consistent with his having been alerted to the subject either by some kind of briefing on Project Feed Back, or by the briefing on the TCP Report which he received in November 1954, or else by a conversation with John Dunning about the Grosse Report in the winter of 1953–4.[2] He also stated that by November 1956 the intelligence services had advised him that the Russians would be able to launch a satellite after another year.[3] That is reasonably consistent with the evidence reviewed in Chapter 10, except for Stewart Alsop's unsubstantiated claim that the first such warning came three years earlier.[4]

Not surprisingly, Eisenhower drew attention to the record of the Truman administration, comparing their total expenditure of only $7 million on developing long-range ballistic missiles with the $178 million spent in the first three years of his administration on the Atlas project alone. He also recalled that at the time of the sputniks crisis he had abstained from making such partisan points, as was confirmed in Chapter 1 from a contemporary document.[5]

Eisenhower accepted that the von Braun team could perhaps have orbited a satellite 'considerably before the Soviets', but pointed out that 'when this capability was discovered in the middle of 1956, the Defense Department and the National Science Foundation showed little inclination . . . to drop Vanguard'.[6] That statement was also confirmed from documentary evidence in Chapter 9.

One point emphasized by Eisenhower which has not been dealt with directly above, and which is never mentioned by space historians, is that in the spring and summer of 1957 the Congress, with both Houses under Democratic Party control, was seeking to impose very substantial cuts on his projected budget, the largest peacetime request in the history of the United States. According to Eisenhower the threatened cuts in the defense budget caused Secretary Wilson to be cautious about using emergency funds to bale out the mushrooming costs of Vanguard.[7]

Eisenhower was least convincing when it came to the rationale behind the selection of the Navy's Vanguard launching system. Although long-range ballistic missiles had been singled out for top priority – a decision which few of the administration's critics, except for Lieutenant General Gavin, ever questioned directly – Eisenhower's claim that in the summer of 1955 the von Braun team was too deeply involved with missile development to take on the satellite is not borne out by the evidence. Eisenhower also argued that, because 'our scientists deliberately planned to share all information acquired with participating scientists all over the world', the satellite programme had had to be separated from missiles so that 'no secret missile information would be involved'.[8] That was disingenuous, for as was shown in Chapter 8 the IGY scientists were expected to share only the information gained *from* satellites, not that which went into the construction and guidance of their rocket launchers. Thus in NSC 5520, and in various IGY discussions prior to October 1957, the fact that launchers would have to be provided by the military was not regarded as detracting from the humane and scientific character of the enterprise, any more than in the case of the vertical sounding rockets used by several national IGY committees during the Year.

Lastly, Eisenhower mentioned the idea that Sputnik 1 had resolved the 'freedom of space' issue, but only as a consolation point raised by Deputy Secretary of Defense Quarles *after* the event (see section 6 below).[9]

*        *          *

Certain points which Eisenhower did not make in his memoirs can be added on his behalf. First, the administration was commendably prompt in accepting the satellite proposal as laid before it in the spring of 1955, and even doubled the amount of money requested. Next, Project Vanguard was given a priority status only just below the top priority reserved for ICBMs and IRBMs. No other part of the IGY and no other American research project of the day enjoyed anything like such official support. Third, the administration was well aware, from the outset, of the implications of the project for America's international prestige. (Its mistake, perhaps, lay in falling in too readily with the scientists' élitist conception of how best to secure that prestige.) Fourth, the same tendency to defer to those pretending to expertise in a field where all were still novices appears to have prevented the administration from insisting firmly enough that the repeated changes of plan which were a feature of Project Vanguard should be thoroughly explained and justified to those who carried the political, if not the technical, responsibility for its success. Rabi's comment about Eisenhower having been modest to a fault appears to contain at least a germ of truth.[10]

Lastly, if priority with a *trackable* satellite was the minimum achievement necessary to sustain American prestige, then the progress of the tracking systems was just as important as that of the launchers. Whether or not the von Braun team had developed the rocketry and guidance capability

to place an Orbiter-style mini-satellite in orbit by mid-1957, they certainly
did not have the optical tracking system which had been a part of their
original proposal. In fact, the first Baker-Nunn satellite-tracking camera
was not even ready by 4 October 1957.[11] On the radio side, two Microlock
stations were built in 1956, one in California and the other at Cape Canaveral. No more were built until after the first sputniks. But the first two did
constitute a basic tracking capability, though one which was not notified
to the administration and was not taken into account at Project Vanguard,
having been developed outside it and for a rival military service. The
primary Minitrack system was barely completed by September 1957, and
final installation and check-out of all the communication circuits linking its
Washington computing centre with ground stations in South America was
not completed until 1 October 1957, three days before Sputnik 1.[12] Even
if the Americans had radically increased their efforts and had included an
early attempt with the Army satellite-launch system, it remains uncertain
how much, if any, lead they might have secured over their Soviet rivals,
particularly as the latter might conceivably have responded to such an
American move (which could not have been kept secret) by reversing
the order of their applications tests for the SS-6 booster and firing satellite
shots before instead of after the ICBM test flights.


3   THE VIEW FROM HUNTSVILLE

The saying that history is written by the victors has a certain truth in the
case of the American space programme. Project Vanguard was closed
down soon after the IGY, although some of its system components were
bequeathed to later more successful satellite launchers. By contrast the
Army Ballistic Missiles Agency, which launched America's first satellite,
was re-organized as NASA's Marshall Space Flight Center with von Braun
as its director. Before he died in 1977 he had led the effort to build the
mighty Saturn rocket boosters which took American astronauts to the
Moon in Project Apollo. Both von Braun himself and several other active
participants in the American space programme have contributed to the
historical literature. The most prolific and distinguished such scholar is
Frederick Ordway, a former head of public information at ABMA and
the Marshall Center, co-author of historical works with von Braun, and
a contributor to the field of astronautics in his own right. The notion
of a 'Huntsville school' of American space historians can reasonably be
stretched to include the former president of the IAF and advocate of Project
Orbiter Frederick Durant, and even perhaps Willy Ley, who shared many

of the same views but seldom if ever collaborated with any of the others. The school can be seen as expressing the viewpoint of the rocket engineers, and they have certainly championed the view of American space policy before the sputniks which portrays von Braun as an ill-treated but dogged pioneer who eventually triumphed against the odds. On the other hand their work has been criticized for a tendency to neglect the social and political aspects of space programmes as if 'to string well known facts together into unimaginative chronicles was sufficient'.[13]

The authors in this group have noted that most of the Truman years were 'years of frustration and disappointment' for von Braun and his fellow veterans of Peenemünde, and they tend to believe that anti-German chauvinism was a major factor in the displacement of Orbiter by Vanguard in 1955.[14] Their low opinion of Eisenhower's grasp of the political significance of space exploration has probably influenced many younger historians and writers on space affairs, through the historical meetings and publications of such bodies as the American Astronautical Society and the IAF.[15] But in print Ordway at least has confined himself to occasional guarded comments on the early Eisenhower period, such as that before the sputniks 'no one regarded the [satellite] efforts as a race to space – at least, the United States government chose not to view its project in those terms', or that (even by 1985) the reasons for the selection of the Vanguard system 'are still not properly understood'.[16] Typically, such historians have difficulty accepting as fact Eisenhower's well-documented enthusiasm, in early 1955, for the Disney/Von Braun film, *Man In Space*.[17] Their judgement remains untouched by the declassification, fourteen years ago, of most of the records of the OCB Working Group that was established to oversee the prestige and propaganda aspects of Project Vanguard.


## 4   THE SCIENCE ADVISERS' VERSION

A distinguished historian of American space policy has argued, on the basis of Eisenhower's consultations immediately after Sputnik 1, that science advice on space was dominated by people who lacked the necessary 'space background'.[18] The shortcomings in the advice put forward before the sputniks by the administration's senior science official, Alan Waterman, were suggested in Chapter 9. But it should not be overlooked that the early space scientists in the US IGY Committee, and those advising the government through the Stewart Committee, usually succeeded in having their recommendations adopted. When for once they failed, over the number of Vanguard launches, they used the test spheres to increase the number anyway. They therefore played a major part in forming the space policy of the Eisenhower administration before the sputniks.

The scientists themselves have been slow to accept any part of the responsibility for the sputniks crisis. Clifford Furnas blamed interservice rivalry, a sentiment echoed by Fred Singer today.[19] Whipple, Van Allen and Singer all believe that political intervention was a major hindrance to American progress.[20] Apart from Furnas, who was on the government side of the relationship for most of the period concerned, none of the scientists and engineers who gave advice and technical briefings about satellites to the administration between 1954 and 1957 has come forward in print to give a full account of what actually took place.[21] They were certainly never asked to do so for Johnson's Preparedness Subcommittee hearings, and other Congressional bodies rarely probed any further.

There is much that surviving science advisers could now help to explain. They seem to have regarded the IGY as, to some extent, a venture contributing to the alleviation of the Cold War (on Western terms), but to have been relatively unconcerned over whether their elected government understood or shared that purpose. Acting as private citizens and without consulting Washington, they drew up the rules for the international transfer of information within the IGY. They arranged for a suitably 'international' suggestion about the desirability of scientific satellites to be made, and then committed their country to make the attempt within a limited time, again without consulting their government. When they started lobbying government for an IGY satellite project in the spring of 1955, they benefited from a tacit association or at the least a confusion of their proposal with Project Orbiter, for which the technical and budgetary requirements were extremely modest. After a vigorous debate a majority on the Stewart Committee recommended that Project Vanguard, a far more risky and expensive, but also more satisfactorily 'scientific' proposal, should become the American entry for the IGY. In defending this decision against repeated Army objections the Stewart Committee and science officials in the Department of Defense effectively concealed from the administration the advent, in 1956, of the first US satellite-tracking system, the JPL Microlock, a potentially key component for an early launch capability. Finally, unidentified members of the Vanguard team, apparently at the Naval Research Laboratory, made the adjustment to the test-sphere schedule that has already been referred to. The upshot was that American humiliation in the face of the first two sputniks was compounded by the very public failure of what had in effect become the United States' first satellite launch attempt on 6 December 1957.

Throughout this period the scientists were not content to stick to the task of designing and launching an Earth satellite, but became embroiled in the secret policy processes which coordinated the cold-war activities of the United States, including the propaganda to be made out of scientific and technical achievements. Yet they have described the IGY, both then and since, as a humane and neutral enterprise, devoted exclusively to the pursuit of scientific truth.

## 5   THE NASA VERSION

NASA was quick to see the importance of historical work, at least with respect to its own activities and those of its predecessor the NACA. Since Eugene Emme was appointed as the first Historian in 1959 the NASA History Office has sponsored an impressive body of work. Although it has some of the limitations of official history it is none the less amongst the best of its kind. But the scholars concerned have seldom been neutral about their patron. Instead, there has been a tendency to assume without question that American space policy languished in a dark age before 1958. However understandable this sentiment may be, it has not always made for objectivity in respect of Eisenhower's early space policy.

The problem has been mitigated by an intellectual 'house style' which emphasizes the production of a clear empirical account of the process of technology development, a by no means easy feat. Most contributors to the NASA History Project have therefore eschewed the invidious task of allocating responsibility for the outcome of the first space race between the various parties to the early development of United States space policy. Indeed, the authors of the official history of Project Vanguard, Constance Green and Milton Lomask, were so anxious to avoid passing judgements of this kind that they ended in self-contradiction. Having first dismissed the question 'might not long-term progress have been faster if the National Security Council had assigned the program top priority and the Department of Defense had then authorized production of the vehicle on a crash basis . . . ?' as one of the 'speculations [that] tend to becloud judgments on Project Vanguard', they then re-introduced the self-same opinion in their discussion of the issue, as coming from 'well-informed people', and forthwith endorsed it as their own:

> In the opinion of well-informed people, a major handicap that beset Project Vanguard from the start was its relegation to a status secondary to the ballistic missile program. The White House decree that

procurement and testing of the satellite vehicle must not interfere with top-priority military projects left the [Naval Research] Laboratory and its prime contractor [Martin Co.] more than once in an awkward position; certainly Vanguard's low priority slowed the flow of money and delayed progress on the vehicle in less obvious ways.[22]

Other NASA historians have sometimes echoed the views of the Huntsville school. Charles Alexander referred to 'the apparent complacency of official Washington' as having continued even after the first two sputniks, and twenty years later Elizabeth Muenger explained the sputniks crisis as arising from the 'low priority' given to Project Vanguard, and from the policy that 'the United States was not in a race with the Soviet Union for the first launch'.[23] But the widely-used expression 'low priority' appears to mean only that Vanguard did not enjoy the same *top* priority with the ICBM and IRBM projects, and the other assertion has never been supported with primary evidence from before the sputniks. As was shown in Chapter 11, there may have been a provisional *public-relations* policy of 'no race' from about June 1957, but there is no evidence that there was ever a corresponding *action* policy.

## 6   SOME INDEPENDENT VIEWS

If American historians with personal or institutional links to their country's space programme as it has developed since 1958 have tended to share the negative feelings of members of the space professions about what occurred before that date, they have also been supported by independent scholars, some of whom have looked more closely into early American space policy than the groups already discussed in this chapter. Enid Schoettle, for example, endorsed the 'low priority' story by assigning officials such as Quarles and Furnas to 'the lower echelons of the national bureaucracy' and by claiming that 'few citizens' knew anything about Project Vanguard before 4 October 1957.[24] (The actual figure, from a survey carried out in the spring of 1957, is 46 per cent.)[25]

John Logsdon is one of the principal authorities for the view that 'even before 1957 . . . the Eisenhower space policy [included] a low estimate of the political significance of a satellite launch'.[26] He has argued that the pursuit of 'strategic political objectives such as national prestige . . . through space activity' was not accepted as a 'fundamental principle of U.S. space policy' until after Sputnik 1, nor indeed until after the close of the Eisenhower administration.[27]

It is certainly possible to contest the latter part of Logsdon's thesis, on the grounds of the Eisenhower administration's evident concern with enhancing American prestige through achievements in space, such as lunar probes or the primitive communications satellites Score and Echo 1, during its last three years in office. But to do so would take us beyond the framework of the present investigation. More recently, Logsdon has revised his position by claiming only that 'the Eisenhower administration explicitly rejected the idea of using *large* space technology projects to compete in symbolic, prestige-oriented accomplishments with the Soviet Union'.[28] If this is meant to apply to Eisenhower's policy before as well as after the sputniks, then on the assumption that not all space projects are 'large', since the term would otherwise be redundant, Logsdon's statement does not rule out the possibility that *some* Eisenhower space projects *were* aimed at upholding the prestige of the United States. The evidence reviewed in Chapter 11 suggests that such was indeed the case with Project Vanguard.

\*        \*        \*

Walter McDougall's 'political history of the space age' has been mentioned frequently in these pages.[29] It presents an over-arching historical thesis, which cannot be summarized here, to the effect that the differing responses of the Soviet Union and the United States to the possibilities opened up by rocketry in the post-war period – a difference which led to the early Soviet priority in space – were aspects of a more fundamental divergence between the ways in which the two societies apprehended and managed their material technologies. In the course of the argument much documentary material which has seldom been handled by previous historians is revealed and explored. McDougall's treatment of the subject has not been greeted with universal acclaim, particularly by those who do not want the question of who in the United States was to blame for the humiliating priority of the sputniks to be answered with 'Everyone concerned!' But the book is an indisputable landmark of scholarship which has finally secured for space history a place within mainstream political historiography.

Alone amongst American space historians, McDougall returns a verdict on Eisenhower's early space policy which is largely positive in tone:

Occupied by the need to keep abreast of the USSR in long-range rocketry, the Eisenhower administration put the ICBM on a crash basis. Absorbed by the need to monitor Soviet R&D and deployment whether arms race or arms control obtained, it also gave priority to the USAF spy satellite program, two and one-half years before the Space Age opened. Worried about the legal and political delicacy of satellite overflight, it seized the IGY opportunity to initiate an unobtrusive scientific satellite program under civilian auspices. Finally, the administration was advised of the propagandistic value of being first into space. Of all these critical policy areas, however, the last had the lowest priority. For there were two ways the legal path could be cleared for reconnaissance satellites. One was if the United States got away with an initial small satellite orbiting above the nations of the earth 'for the advancement of science' – and had no one object to it. The other way was if the Soviet Union launched first.

The second solution was less desirable, but it was not worth taking every measure to prevent.[30]

There are several difficulties with the intriguing claim that the Eisenhower administration was so concerned about the legal problem of satellite overflights that officials became reconciled to the 'second solution' even *before* they were obliged to make the best they could of the sputniks as a *fait accompli*. The first is that, although the State Department did consider this problem during 1954 and 1955, and although such considerations entered into the National Security Council discussions which resulted in the approval of NSC 5520,[31] McDougall has provided neither documentary evidence nor recollections of protagonists to the effect that the particular line of reasoning which he outlines was prominent, or even present, in those discussions. The second is that although the OCB Working Group was charged amongst other things with handling the legal aspects of the satellite project there is only one brief and oblique reference to the 'freedom of space' problem in its minutes between July 1955 and 4 October 1957, and none at all to McDougall's hypothetical 'second solution' to it.[32] Lastly, the 'peaceful' and 'scientific' satellite overflights of the IGY did not in fact resolve the legal difficulty when it came to overflights by secret military satellites. The Soviet Union voiced numerous objections to such overflights in the late 1950s and early 1960s, and the only development which in practice resolved the issue was the Soviet acquisition of similar space technology, a few years after it was deployed by the United States.

Before Sputnik 1 there was a developing consensus about the probable legal position for future artificial satellites, which was confirmed by the experience of the IGY. Nations might well concede a *de facto* right of passage to peaceful satellites, engaged on missions of general humanitarian value such as scientific research, but they would probably hesitate to extend such rights to other satellites not so engaged.[33] After 1957 Soviet jurists insisted that the legality of the first satellites had stemmed from the framework of international agreement within which they had been planned and launched, and argued that the best basis for establishing the 'freedom of space' would be an international agreement banning all military uses of space.[34] When it came to intelligence-gathering satellites:

> From the viewpoint of the security of a state it makes absolutely no difference from what altitude espionage over its territory is conducted. A state will not feel any safer because military preparations against it are carried on at a very high altitude. . . . Hence there is absolutely no ground for alleging that espionage at a high altitude, with the aid of artificial Earth satellites, is quite lawful. . . . Any attempt to use satellites for espionage is . . . unlawful.[35]

There is therefore no primary evidence for McDougall's hypothesis about Eisenhower's pre-sputnik space policy. No such legalization for reconnaissance satellites was in fact achieved as a result of the Soviet 'first' in space, nor could one have been reasonably anticipated by American policy-makers. For these reasons, McDougall's hypothesis, however ingenious, must be rejected.


# 7 CONCLUSIONS

Three conclusions are supported by the re-examination of the origins and conduct of American space policy between 1945 and 1957 which has been conducted here. First, that the Truman administration has a considerable share in the historical responsibility for the sputniks crisis, even though it left office over four years before it. Second, that in addition to President Eisenhower and his leading officials many other people had a part in determining the conduct of the first public American space project, Project Vanguard. And third, that the historiographical tradition of discounting these factors in early American space policy is probably due to the lasting influence of the biased or ill-informed indictments of the Eisenhower administration that were pronounced in the press and at Congressional hearings immediately after the first sputniks. The

President and a small group of his senior officials made an irresistibly convenient scapegoat for an informal alliance of scientists, engineers, businessmen, publicists, and rival politicians, all of whom were concerned to establish and protect interests in the field of space exploration for a future extending beyond the closing years of the Eisenhower administration. Above all, the Democratic hawks required a matching 'space gap' to support their 'missile gap' campaign, a 'space gap' which had to be blamed largely on the negligence of the incumbent administration, with occasional accusations of Soviet bad faith thrown in to complete the picture.

The research which has been reported in this book reveals that the Eisenhower administration was far more concerned about the political, and above all the prestige aspects of its IGY satellite project, before the sputniks, than has previously been supposed. It also shows that the IGY arrangements were far more confused and the Russian scientists more open and cooperative than is generally recognized.

However, although this study has emphasized the positive aspects of Eisenhower's early space policy and the role of other parties in bringing about the sputniks crisis, it would be wrong to seek to overturn completely the traditional attribution of blame for America's humiliation in the fall of 1957. Eisenhower and his colleagues in the National Security Council clearly failed to communicate the policy laid down in NSC 5520 effectively to all who would be concerned, in practice, with its implementation. No official or group of officials at the White House was assigned to liaise closely and directly with those tasked with the actual execution of Project Vanguard, rather than its political presentation. General Goodpaster might from time to time seek to inform himself and the President, but he had no way of telling whether the answers he received from the Department of Defense or the National Science Foundation were reliable. In particular, Vanguard's serious budgetary problems, having been noticed by senior officials as early as January 1956, could surely have been remedied sooner than they were. Also, more could have been done to inform the American people of the great strides being made by the Soviet rocket programme, with its implications for both ICBMs and satellites, though in a back-handed way the administration's reticence on this matter before the sputniks, and its attempts to curb any public references to it by IGY scientists, do show that it was by no means indifferent as to which side would come first.[36] (What it felt obliged to *say* later, when faced with the Soviet victory, is of course irrelevant to the argument about what its policy had previously been.)

Finally, two or three senior officials, no doubt under great stress, foolishly exacerbated the administration's public-relations problem after Sputnik 1 by letting slip derogatory remarks about the Soviet achievement. But even here historians have preferred simply recalling the unfortunate comment of Eisenhower's senior aide, Sherman Adams, about not wishing to win 'high score in a celestial basketball game', to reminding their readers that the basketball metaphor which the administration's critics affected to find so deeply shocking had in fact been widely promoted by the IGY satellite scientists themselves, and by the information media, for over two years before Adams blundered into his quip.[37]

\*     \*     \*

The ultimate test for any body of historical work lies in the reflection of itself which is returned from more popular, tertiary treatments of its field. By that criterion, the historiography of early American space policy still leaves much to be desired. A few examples only must suffice. Richard Lewis, a journalist who covered the Apollo programme and then wrote books about American space activities, simply antedated the National Academy's first approach to government by five months and commented that 'The President took no action on it.' He was then able to suggest that it was only after news of the formation of the Soviet Academy's space commission (ICIC) in April 1955 that 'the political merit' of the satellite idea succeeded in 'penetrat[ing] the reserve at the White House toward space projects'.[38] With a few strokes of the pen the two-month process of largely positive inter-departmental consultations, which led from Waterman's actual approach in mid-March 1955 to the NSC's endorsement of an IGY satellite project in May, had been transformed into a seven-month struggle to overcome a short-sighted and obstructive President.

In 1982 the British author and journalist Andrew Wilson explained the Stewart Committee's selection of Vanguard on the grounds that 'Beating the Soviets was not so important in 1955 – except perhaps to the military.'[39] Such a comment might not have been made if the historical literature had paid any attention to Nelson Rockefeller's strongly-worded appendix to NSC 5520, which stated that 'The stake of prestige that is involved makes this a race that we cannot afford to lose.'[40] Wilson might also have written differently if any American historian had drawn his attention to the existence and work of the OCB Working Group on NSC 5520.

The British science writer, Tony Osman, can furnish a final example from the tertiary literature. According to Osman, Defense Secretary Wilson and President Eisenhower were actually 'opposed to research into powerful rockets'.[41] This remark was not accompanied by an explanation of how, in that case, the Atlas, Titan, Thor, Jupiter and Polaris missile projects, together with most American space launch systems used before the Shuttle, including the Saturn booster used in the Apollo Project, were all initiated under the Eisenhower administration.

But it is only necessary to recall the example given in the Preface to suggest that the inadequacy of such sketches of Eisenhower's early space policy should not be blamed on popular writers who rely in good faith on an impressive corpus of historical scholarship, so much as on the scholars themselves who have been sustaining one of the Cold War's favourite myths for over thirty years, and who in doing so have been serving distinctly short measure to those who are obliged to depend upon them.

# Notes and References

Books and journal articles are cited in full bibliographical detail only at the first reference, to which later citations refer back directly or indirectly. The dates of British editions, which have sometimes been used for citations, are often different from those of American editions discussed in the text. News reports in newspapers and magazines are usually cited by date only.

Committees of the US Congress are usually referred to by abbreviations of which only 'Astronautics' for the House Select Committee on Astronautics and Space Exploration and its successor the House Committee on Science and Astronautics, and 'Space Sciences' for the Senate Special Committee on Space and Astronautics and its successor the Senate Committee on Aeronautical and Space Sciences, require explanation. Subcommittee details are given only at the first citation. Official United States publications were published by the US Government Printing Office, Washington, unless otherwise stated.

Document collections, whether published or unpublished, are referred to by the abbreviations listed below, followed by such further information as is required to locate the document. Where it has been necessary to refer to a document via a secondary source details of the latter are also given.

Standard postal abbreviations are used for American states when appropriate. Other abbreviations not listed below are included in the list of abbreviations and technical terms on p. xiii.

| | |
|---|---|
| BCIGY | Records of the British Committee for the IGY, Royal Society, London. |
| COHC | Oral History Collection of Columbia University. |
| DD-[year] | *Declassified Documents* (Washington: Carrollton Press, then Research Publications). |
| DDEL | Dwight D. Eisenhower Library, Abilene, Kansas. |
| DeHaven | Document collection attached to: E. M. DeHaven, *The Evolution of USAF Weapons Acquisition Policy, 1945–1961* (Andrews Air Force Base, MD: USAF Historical Office, 1962). |
| HSTL | Harry S. Truman Library, Independence, Missouri. |
| LBJL | Lyndon Baines Johnson Library, Austin, Texas. |
| NAS | US National Academy of Sciences, Central Files. |
| NHO | NASA History Office, Washington. |
| OH Interview | Transcript of Oral History Interview. |
| OSANSA | Office of the Special Assistant for National Security Affairs. |

| | |
|---|---|
| OSAST | Office of the Special Assistant for Science and Technology. |
| *PPPUS-[year]* | *Public Papers of the Presidents of the United States: Dwight D. Eisenhower* (Washington: Office of the Federal Register, National Archives and Records Service, 1958–60). |
| RG | Record Group at USNA. |
| SAOHP | Space Astronomy Oral History Project, National Air and Space Museum, Smithsonian Institution, Washington. |
| Secretary | Office of the Staff Secretary. |
| TPESP | Papers of the USNC-IGY Technical Panel on the Earth Satellite Project. |
| USNA | US National Archives, Washington. |
| USNC-IGY | Papers of the US National Committee for the IGY, National Academy of Sciences, Washington. |
| WHO | White House Office Files. |

The author was privileged to interview some of the surviving protagonists of the events discussed in this book. The dates of those conversations, several of which are quoted or cited as 'interview' in these Notes, are listed here. All except those marked with an asterisk were tape-recorded. The author records his gratitude for the original interviews, and in several cases also for copies of papers and subsequent correspondence, to: Frederick C. Durant, III (5 June 1987); Eilene M. Galloway (11 April 1987); George W. Hoover (11 May 1987); James R. Killian (dec.) (13 June 1987); Ward Kimball* (12 May 1987); Marcel Nicolet* (27–8 September 1989); Frederick I. Ordway, III (5 June 1987); George E. Reedy (15 May 1987); Milton W. Rosen (2 June 1987); S. Fred Singer (24 May 1987); Kurt R. Stehling (24 April 1987); James A. Van Allen (18 May 1987); Fred L. Whipple (11 June 1987); Glen P. Wilson (20 April 1987); Herbert F. York (13 May 1987).

### Preface

1. P. N. James, *Soviet Conquest from Space* (New Rochelle, NY: Arlington House, 1974); P. Schafly and C. Ward, *Strike from Space* (Alton, IL: Pere Marquette, 1965).
2. W. W. Rostow, *The United States in the World Arena: an Essay in Recent History* (New York: Harper & Row, 1960), p. 356. Many people mentioned in this book were or still are holders of the academic doctorate, professorships, or membership in national scientific academies. For simplicity's sake such titles have been omitted unless directly relevant to the narrative.

3.  C. R. Koppes, *JPL and the American Space Program: a History of the Jet Propulsion Laboratory* (New Haven, CT: Yale U. Press, 1982), pp. ix, 40–1.

## Chapter 1: A Devastating Blow

1.  This is the time currently accepted by Western experts: D. G. King-Hele, J. A. Pilkington, D. M. C. Walker, H. Hiller and A. N. Winterbottom, *The RAE Table of Earth Satellites 1957–82* 2nd edn (London: Macmillan, 1983), p. 1. It also agrees with the best anecdotal evidence about the launch from Russian sources. Earlier estimates put the launch time slightly later: ibid., p. ii; or earlier: K. W. Gatland, *Astronautics in the Sixties* (London: Iliffe Books, 1962), p. 139.
2.  J. A. Van Allen, *Origins of Magnetospheric Physics* (Washington: Smithsonian Institution, 1983), p. 46.
3.  *New York Times*, 5 October 1957.
4.  NHO – American Rocket Society Files, 'Proc. 1st Meeting of Ad Hoc Committee on Space Flight', 17 May 1952.
5.  *New York Times*, 6 October 1957.
6.  Ibid.
7.  Ibid.
8.  *New York Times*, 15 October 1957; New York *Herald Tribune*, 16 October 1957.
9.  G. E. Reedy, *The Twilight of the Presidency* (New York: World Publishing, 1970), p. 54.
10. Supporters of the Republican Party called in vain for a truly non-partisan response to the Soviet achievement. On 17 October, for example, the playwright and diplomat Clare Boothe Luce told the A. E. Smith Memorial Dinner that 'It is twelve years from Hiroshima to the moon [i.e. Sputnik 1]. Six of them [actually seven and a half] were Democratic, six [four and three-quarters] Republican. Let us say no more.' *New York Times*, 20 October 1957.
11. US Congress, Senate, Armed Services, Preparedness Investigating Subcommittee, Hearings: *Inquiry into Satellite and Missile Programs* (85th Congress, 2nd Session, 1958 – hereafter *ISMP*).
12. New York *Herald Tribune*, 16 October 1957.
13. US Congress, Senate, Armed Services, Hearings: *Study of Airpower* (84th Congress, 2nd Session, 1956), p. 1141.
14. C. C. Furnas, 'Why Did U.S. Lose the Race?' *Life*, 21 October 1957.
15. *Time*, 21 October 1957.
16. *New York Times*, 20 October 1957.
17. *New York Times*, 31 October 1957.
18. *New York Times*, 1 November 1957.

19.  *New York Times*, 5 November 1957. Long before the controversy aroused by the sputniks *Time* (30 January 1956) had published a balanced assessment of Keller's contribution to the American missiles programme, including his neglect of the ICBM.

20.  *New York Times*, 10 and 24 November, 10 and 12 December 1957.

21.  *New York Times*, 10 December 1957.

22.  *New York Times*, 12 December 1957.

23.  US Congress, House, Appropriations, Subcommittee on War Department Appropriations, Hearings: *Military Establishment Appropriation Bill for 1947* (79th Congress, 2nd Session, 1946), p. 1117.

24.  *New York Times*, 10 November 1957.

25.  *New York Times*, 10 December 1957.

26.  U. Albrecht, 'The Nuclear-Propelled Bomber' in H. G. Brauch (ed.), *Military Technology, Armaments Dynamics and Disarmament* (London: Macmillan, 1989).

27.  *ISMP* (n. 11), pp. 582–5. On the issue of the late American start von Braun was supported at the hearings by George Clement, who had worked on satellite studies at the RAND Corporation for over ten years: *ISMP* (n. 11), p. 2139. Even Edward Teller, who was chosen by Johnson's staff to spearhead the attack on the administration, told the Subcommittee that the American ICBM programme had been started 'too late': *ISMP* (n. 11), pp. 6, 9. He was not invited to explain the remark.

28.  DDEL – Harlow Papers, Box 2, Folder: Missiles, Misc. 1957–58 (2), W. F. Knowland and J. W. Martin, 'A Public Memorandum', 13 January 1958. See also ibid., Folder: Missiles, Misc. 1957–58 (1), 'Fiscal Year 1947 Funds Impounded by Truman', February 1958, and 'Here's What President Truman Was Told After World War II About Rockets and Missiles', 18 February 1958.

29.  Knowland and Martin, 'Public Memorandum' (n. 28). Comparative figures for missile expenditures under the Truman and Eisenhower administrations were first published by the Republican National Committee in November 1957: *Washington Daily News*, 11 December 1957. See also: D. W. Cox and M. Stoiko, *Spacepower: What It Means To You* (Philadelphia: John C. Winston, 1958), p. 44.

30.  *Washington Post*, 13 December 1957.

31.  DDEL – WHO: Secretary, Subjects – Alphabetical, Folder: S.A.C. (2), Gen. A. J. Goodpaster, 'Memorandum of Conference with the President, October 15, 1957, 11 AM', pp. 3–4.

32.  The best recent account of the Soviet space programme from its origins to the present day is probably: B. Harvey, *Race Into Space: the Soviet Space Programme* (Chichester: Ellis Horwood, 1988).

33.  R. Witkin (ed.), *The Challenge of the Sputniks* (Garden City, NY: Doubleday, 1958).

34.  J. M. Gavin, *War and Peace in the Space Age* (New York: Harper & Brothers, 1958; London: Hutchinson, 1959).

35. E. Bergaust and S. Hull, *Rocket to the Moon* (Princeton: Van Nostrand, 1958).

36. Editorial *Missiles and Rockets*, November 1957 (repr. as advertisement *New York Times*, 7 November 1957).

37. Cox and Stoiko, *Spacepower* (n. 29), p. 30.

38. C. C. Adams, *Space Flight: Satellites, Space Ships, Space Stations, and Space Travel Explained* (New York: McGraw-Hill, 1958), p. 57 – emphasis added.

39. Cox and Stoiko, *Spacepower* (n. 29), pp. 30, 76; W. Lippmann, 'The Portent of the Moon', New York *Herald Tribune*, 10 October 1957.

40. D. Pearson and J. Anderson, *USA – Second-Class Power?* (New York: Simon and Schuster, 1958), p. 252.

41. D. Caute, *The Great Fear: the Anti-Communist Purge under Truman and Eisenhower* (London: Secker & Warburg, 1978), p. 480.

42. Pearson and Anderson, *USA* (n. 40), pp. 57–8; c.f. DDEL – Harlow Papers, Box 2, Folder: Missiles, Misc. 1957–58 (2), unsigned memo. (probably P. Areeda for Harlow), December 1957: 'Symington is complaining about the attacks on him by Meade Alcorn, Jerry Ford etc. He says if it doesn't stop he will attack the President as being responsible in 1947 for the cancellation of the MX-774 ICBM.' The allegation that Eisenhower had personally 'killed' the American ICBM in 1947 was repeated in the space literature for several years. See for example: D. W. Cox, *The Space Race: from Sputnik to Apollo . . . and Beyond!* (Philadelphia: Chilton Books, 1962), p. 20.

43. Pearson and Anderson, *USA* (n. 40), p. 66.

44. A. C. Ivy, 'Space Physiology – Physiological Man Versus His Technological Twin' in US Congress, Senate, Space Sciences, Committee Print: *Compilation of Materials on Space and Astronautics No. 2* (85th Congress, 2nd Session 1958), pp. 169–70. The unnamed participant in the symposium referred to was probably Ivy himself.

45. W. von Braun and F. I. Ordway, *History of Rocketry & Space Travel* (London: T. K. Nelson, 1967), p. 179.

46. W. von Braun, F. I. Ordway and D. Dooling, *Space Travel: a History* (New York: Harper & Row, 1985), p. 170.

47. W. A. McDougall, . . . *the Heavens and the Earth: a Political History of the Space Age* (New York: Basic Books, 1985), p. 97.

48. R. I. P. Bulkeley and G. Spinardi, *Space Weapons: Deterrence or Delusion?* (Cambridge: Polity Press, 1986).

**Chapter 2: *Ad Homines* – Eisenhower and Johnson Before 1957**

1. D. D. Eisenhower, *Crusade in Europe* (London: Heinemann, 1948), pp. 284–5.

2. W. F. Craven and J. L. Cate, *The Army Air Forces in World War II* vol. 3 (Chicago: Chicago U. Press, 1951), pp. 102–3, 527, 879.

3. B. Collier, *The Battle of the V Weapons 1944–45* (Morley: Elmfield Press, 1976), p. 111.

4. Ibid., p. 129.

5. T. Bower, *The Paperclip Conspiracy* (London: Michael Joseph, 1987), pp. 3, 134.

6. Ibid., p. 255.

7. USNA – RG 319, Army Intelligence Decimal File 1941–48, Box 1001, 400.112 (230.312), 6 November 1948, quoted in: Bower, *Conspiracy* (n. 5), p. 284.

8. US Congress, House, Appropriations, Hearings: *Military Appropriation for 1947* (1946 – ch. 1, n. 23), p. 1117.

9. US Congress, House, Appropriations, Subcommittee on War Department Appropriations, Hearings: *Military Establishment Appropriation Bill for 1948* (80th Congress, 1st Session, 1947), p. 79.

10. Ibid., p. 78.

11. DDEL – Pre-Presidential Papers: Subjects, Box 145, carbon typescript of Eisenhower's testimony to a Joint Executive Session of the House Armed Services and Foreign Affairs Committees, 2 February 1951.

12. S. E. Ambrose, *Eisenhower* vol. 1 (London: Allen & Unwin, 1984), p. 506.

13. DD-1975 – 159A: Ltr, Eisenhower to Collins, 27 June 1951.

14. H. L. Goodwin, *The Science Book of Space Travel* (New York: Franklin Watts, 1954), p. 3.

15. B. W. Cook, *The Declassified Eisenhower* (Garden City, NY: Doubleday, 1981), pp. 13–15.

16. Ibid., pp. 126–33; J. Hale, *Radio Power, Propaganda, and International Broadcasting* (London: Paul Elek, 1975).

17. DDEL – Pre-Presidential Papers: Subjects, Box 145, Folder: Hearings, 1951 (1), Memo. E. W. Barrett for Lt. Gen. A. M. Gruenther, 'Psychological Exploitation of Assumption by General Eisenhower of Command of Supreme HQ Allied Powers in Europe', 3 January 1951.

18. DDEL – OH Interview with L. Norstad, quoted in: Ambrose, *Eisenhower* (n. 12), p. 501.

19. Ltr, Eisenhower to Lovett, 25 September 1951, quoted in: Ambrose, *Eisenhower* (n. 12), p. 503.

20. DDEL – Pre-Presidential Papers: Subjects, Box 145, Folder: Hearings, 1951 (1), 'Testimony to a Joint Executive Session of the Senate Committees on the Armed Services and on Foreign Relations, February 1, 1951', carbon typescript, pp. 8, 75. Words *underlined* were omitted from the published version: US Congress, Senate, Armed Services and Foreign Relations, Hearings: *Assignment of Ground Forces of the U.S. to Duty in the European Area* (82nd Congress, 1st Session, 1951), pp. 10, 28.

21. B. Mooney, *The Lyndon Johnson Story* (London: The Bodley Head, 1964), pp. 28–30.

22. S. L. Rearden, *The Formative Years 1947–1960* part 1 of A. Goldberg

(ed.), *History of the Office of Secretary of Defense* (Washington: OSD History Office, 1984), pp. 353–6.

23. *New York Times*, 14 February 1950.
24. *New York Times*, 10 September 1950.
25. LBJL – Senate Papers: Preparedness Investigating Subcommittees, Box 346, Folder: Correspondence & Memos. 1951–52, Memo. Reedy for Johnson, 22 January 1952.
26. L. B. Johnson, *The Vantage Point: Perspectives of the Presidency 1963–1969* (London: Weidenfeld & Nicholson, 1972), p. 271.
27. L. B. Johnson, 'We Are Ahead of Russia – We Can Stay Ahead', *Preview Magazine*, December 1950.
28. *New York Times*, 4 December 1951.
29. LBJL – Senate Papers: Siegel Papers, Box 407, Folder: Preparedness Investigating Subcommittee – Misc. 1957, 'List of the Major Projects Being Handled by the P.I.S. Staff', 24 January 1957; ibid., Folder: Memoranda – 1957, Memos, Siegel to Johnson, 2 and 7 February 1957.

## Chapter 3: Advice on Missiles for Truman

1. Notable amongst these have been: J. R. Killian, *Sputnik, Scientists, and Eisenhower: a Memoir of the First Special Assistant to the President for Science and Technology* (Cambridge, MA: MIT Press, 1977); G. B. Kistiakowsky, *A Scientist at the White House: the Private Diary of President Eisenhower's Special Assistant for Science and Technology* (Cambridge, MA: Harvard U. Press, 1976); H. F. York, *The Advisors: Oppenheimer, Teller, and the Superbomb* (San Francisco: W. H. Freeman, 1976); D. K. Price, *Government and Science* (New York: New York U. Press, 1954); R. Gilpin, *American Scientists and Nuclear Weapons Policy* (Princeton: Princeton U. Press, 1962); and E. B. Skolnikoff, *Science, Technology, and American Foreign Policy* (Cambridge, MA: MIT Press, 1967). A comprehensive bibliography of the subject is given in the last section of the essay on sources in D. J. Kevles, *The Physicists: the History of a Scientific Community in Modern America* (New York: Vintage, 1979), pp. 457–64.
2. Kevles, *Physicists* (n. 1), pp. 341–66.
3. US Department of State, 'Statement by President Truman at a Meeting at Blair House, Washington, July 14, 1949, 8:15 p.m.' *Foreign Relations of the United States: 1949* vol. 1 (1976), p. 481.
4. Ltr, Oppenheimer to Chairman of the General Board of the Navy, quoted in: Gilpin, *American Scientists* (n. 1), pp. 74–6, as written 'in 1948'.
5. D. A. Rosenberg, 'A Smoking Radiating Ruin at the End of Two Hours', *International Security*, vol. 6, no. 3, 1981/82.
6. US Atomic Energy Commission, Personnel Board, Hearings: *In the*

*Matter of J. Robert Oppenheimer* (1954 – hereafter *IMJRO*), testimony of Luis Alvarez, pp. 773 ff. See also Gilpin, *American Scientists* (n. 1), pp. 73–111; J. Major, *The Oppenheimer Hearing* (London: Batsford, 1971), pp. 97–129.

7.  *New York Times*, 18 and 19 September, 26 October 1951.
8.  Gilpin, *American Scientists* (n. 1), p. 115.
9.  DDEL – Quarles Papers, Box 3, Folder: Trip File (2), list of Project Charles personnel, 28 February 1951.
10. *IMJRO* (n. 6), p. 924.
11. Ibid., p. 935.
12. Ibid., pp. 94–5; see also ibid., pp. 598–600, 923, 936–7.
13. The ONR provided at least part of the funding for more than half of all basic research projects in the physical sciences in the United States between 1945 and 1957, though in the second half of that period the newly created National Science Foundation, headed by former ONR chief scientist Alan Waterman, began to share this role: J. E. Pfeiffer, 'The Office of Naval Research', *Scientific American*, vol. 180, February 1949. Most American participants at the 1953 Oxford conference on upper atmosphere research with rockets, for example, were funded by ONR: R. L. F. Boyd and M. J. Seaton (eds), *Rocket Exploration of the Upper Atmosphere*, supplement to *J. Atmospheric and Terrestrial Physics*, vol. 1, 1954.
14. T. von Kármán (with L. Edson), *The Wind and Beyond: Theodore von Kármán, Pioneer in Aviation and Pathfinder in Space* (Boston: Little, Brown, 1967), p. 326.
15. US Department of State, L. V. Berkner, *Science and Foreign Relations*, report of the International Science Policy Survey Group, DOS Publication No. 3860, May 1950.
16. HSTL – Presidential Papers, Office File, Box 680, Folder: 192A-MISC, Ltr, Comstock to Short, 1 November 1951.
17. Skolnikoff, *Science and Foreign Policy* (n. 1), pp. 252–7.
18. D. Lasser, *The Conquest of Space* (New York: Penguin Press, 1931), p. 262.
19. W. Ley, *Rockets: the Future of Travel Beyond the Stratosphere* 2nd edn (New York: Viking Press, 1944), pp. 271, 253.
20. R. V. Jones, *Most Secret War* (London: Hamish Hamilton, 1978), pp. 459–60.
21. W. Ley, *Rockets: the Future of Travel Beyond the Stratosphere* 3rd edn (New York: Viking, 1945); a revised edition of C. P. Lent, *Rocket Research: History and Handbook* (New York: Pen-Ink Co., 1944 also appeared in January 1945.
22. For example: G. E. Pendray, *The Coming Age of Rocket Power* (New York: Harper & Brothers, 1945). Like Lent, Pendray was a prominent member of the American Rocket Society, and the book is notable for its refusal to discuss the relevance of rocketry to spaceflight, a 'hard-headed' attitude typical of the post-war ARS.

23. B. Brodie, *The Atomic Bomb and American Security*, Memorandum No. 18 (New Haven, CT: Yale Institute of International Studies, 1945), p. 3.

24. *New York Times*, 29 June 1945.

25. *New York Times*, 12 October 1945.

26. B. Brodie, 'War in the Atomic Age' in B. Brodie (ed.), *The Absolute Weapon* (New York: Harcourt, Brace, 1946), pp. 30–2.

27. I. A. Getting, 'Facts About Defense', *The Nation*, 22 December 1945.

28. L. N. Ridenour, 'There Is No Defense' in D. Masters and K. Way (eds), *One World or None* (New York: McGraw-Hill, 1946). A perceptive account of the rapidity and intensity with which Americans applied to themselves the atomic vulnerability which they had brought about, in the first instance, for others, is contained in: P. Boyer, *By the Bomb's Early Light: American Thought and Culture at the Dawn of the Atomic Age* (New York: Pantheon, 1985), pp. 3–26.

29. *New York Times*, 15 February and 21 March 1946.

30. *Life, Time, Newsweek*, 9 December 1946.

31. J. and S. Alsop, 'Are We Ready for a Push-Button War?' *Saturday Evening Post*, 6 September 1947.

32. US Congress, House, Appropriations, Hearings: *Military Appropriation for 1948* (1947 – ch. 2, n. 9), p. 640.

33. F. and K. Drake, 'Our Next Pearl Harbor', *The Atlantic Monthly*, October 1947.

34. Brodie, 'Atomic Age' (n. 26), p. 79 – emphasis added; c.f. Brodie, *Atomic Bomb* (n. 23), p. 13.

35. M. Lehman, *This High Man: the Life of Robert H. Goddard* (New York: Farrar, Strauss, 1963), pp. 169–71, 203, 212.

36. J. V. Becker, *The High-Speed Frontier: Case Histories of Four NACA Programs, 1920–1950* (Washington: NASA, 1980), p. 31.

37. C. R. Koppes, *JPL and the American Space Program* (1982 – Preface, n. 3), pp. 2–77.

38. T. Bower, *The Paperclip Conspiracy* (1987 – ch. 2, n. 5); F. I. Ordway and M. R. Sharpe, *The Rocket Team* (New York: Thomas Y. Crowell, 1979).

39. US Department of Defense, H. H. Arnold, *Third Report of the Commanding General of the U.S. Army Air Forces to the Secretary of War* (1945), pp. 67, 68.

40. von Kármán, *Wind* (n. 14), p. 271.

41. USAF Scientific Advisory Group, *Where We Stand*, 22 August 1945; *Toward New Horizons: Science the Key to Air Supremacy* (33 vols), 15 December 1945; the author has not been able to consult either of these limited-issue publications directly.

42. Alsop and Alsop, 'Push-Button War' (n. 31), p. 102.

43. Von Kármán, *Wind* (n. 14), p. 289.

44. In his memoirs, von Kármán blamed Vannevar Bush for American tardiness in developing long-range ballistic missiles: 'In 1949 and 1950

the Air Force, in keeping with our original report, wanted to proceed with the Atlas on a large scale, but couldn't get enough money from Congress. One source of strength for Congress's rejection, I recall, was that Dr Vannevar Bush . . . ' etc.: *Wind* (n. 14), p. 300. This was refuted by the historian of the American ICBM programme, Edmund Beard, who showed that in 1949 and 1950, *after* Bush had left the RDB, the USAF's Air Materiel Command blocked a suggestion from RAND that the exclusive emphasis on air-breathing missiles should give way to a combined effort in which long-range ballistic rockets would also play a significant part. According to Beard, the AMC 'considered the step-wise development of long-range missiles through the subsonic turbojet Snark and the low supersonic ramjet Navaho to be the most sensible approach. Ballistic rockets were still considered to be far in the future.': E. Beard, *Developing the ICBM: a Study in Bureaucratic Politics* (New York: Columbia U. Press, 1976), p. 96. This policy had first been proposed in *Toward New Horizons*, not over Bush's but over von Kármán's signature.

45.  V. Van Dyke, *Pride and Power: the Rationale of the Space Program* (London: Pall Mall, 1965), p. 10; see also H. L. Dryden, 'Toward the New Horizons of Tomorrow', *Astronautics*, vol. 12, January 1963. Extraordinarily, von Kármán retained his scepticism about ICBMs until at least 1956; in an interview that year he expressed the opinion that ICBMs would not be accurate enough for military purposes, setting the required performance at a 50 per cent probability of striking within five miles of the target: New York *Herald Tribune*, 24 April 1956. But even first-generation ICBMs, deployed only a few years later, either met or came close to that requirement, and within ten years of the interview ICBMs with an accuracy of less than two miles had almost completed their development: J. M. Collins, *US-Soviet Military Balance: concepts and capabilities 1960–1980* (New York: McGraw-Hill, 1980), pp. 443 ff.

46.  US Congress, Senate, Special Committee on Atomic Energy, Hearings: *Atomic Energy* (79th Congress, 2nd Session, 1946), pp. 179–80. For over thirty years American space writers and historians have been requoting the heavily edited and shortened version of this passage which was published soon after the first sputniks – US Congress, Senate, Armed Services, Hearings: *Inquiry into Satellite and Missile Programs* (1958 – ch. 1, n. 11), pp. 822–3. The opportunity has been taken to restore it to the record in its original, slightly less dogmatic, form.

47.  US Congress, Senate, Atomic Energy, *Atomic Energy* (n. 46), pp. 81, 251–2, 283, 396.

48.  W. A. McDougall, . . . *the Heavens and the Earth* (1985 – ch. 1, n. 47), p. 98.

49.  J. L. Chapman, *Atlas: the Story of a Missile* (New York: Harper & Brothers, 1960), p. 56.

50. Beard gives both figures for the number of projects cut at this point: *ICBM* (n. 44), pp. 55, 89.

51. F. J. Malina, 'America's First Long-Range Missile and Space Exploration Program: the ORDCIT Project of the Jet Propulsion Laboratory' in R. C. Hall (ed.), *Essays on the History of Rocketry and Astronautics* (Washington: NASA, 1977), p. 375.

52. DeHaven – Ltr, Maj. Gen. B. W. Chidlaw to Commanding General AAF (C. A. Spaatz), 12 May 1947, quoted in: Beard, *ICBM* (n. 44), p. 56.

53. US President's Advisory Commission on Universal Training, Report, *A Program for National Security*, 29 May 1947, p. 12; B. Brodie and E. M. Galloway, *The Atomic Bomb and the Armed Services* Public Affairs Bulletin No. 55 (Washington: Library of Congress Legislative Reference Service, 1947), pp. 30–1.

54. V. Bush, *Modern Arms and Free Men* (New York: Simon and Schuster, 1949), pp. 84–5.

55. B. Brodie, 'A Critique of Army and Navy Thinking on the Atomic Bomb', *Bull. Atomic Scientists*, vol. 3, no. 8, August 1947; see also H. H. Arnold, 'Air Force in the Atomic Age' in Masters and Way (eds), *One World* (n. 28), p. 29.

56. Bush, *Modern Arms* (n. 54), pp. 85–7.

57. See for example US Congress, House, Armed Services, Hearings: *Investigation of National Defense Missiles* (85th Congress, 2nd Session, 1958), p. 3987.

58. This point was made by one of the US Army's senior R&D officers, Brigadier General Austin Betts, at ibid., p. 4161. For a more detailed explanation of why warhead weight was essentially a 'false issue' from about 1949, see Beard, *ICBM* (n. 44), pp. 140–3.

## Chapter 4: Advice on Space for Truman

1. D. Lang, 'V-2s Among the Rattlers', *The New Yorker*, March 1946; D. Lang, 'What's Up There?' *The New Yorker*, April 1948.

2. L. N. Ridenour, 'Pilot Lights of the Apocalypse', *Fortune*, January 1946.

3. W. von Braun, 'Survey on the Development of Liquid Rockets in Germany and their Future Prospects' in F. Zwicky (ed.), *Report on Certain Phases of War Research in Germany* vol. 1 (Azusa, CA: Aerojet Engineering Corporation), 1 October 1945; repr. *J. Brit. Interplanetary Soc.*, vol. 10, no. 2, March 1951.

4. Project RAND, F. H. Clauser (ed.), *Preliminary Design of an Experimental World-Circling Spaceship*, Report No. SM-11827, 2 May 1946.

5. J. and S. Alsop, 'Are We Ready for a Push-Button War?' (1947 – ch. 3, n. 31) – emphasis added.

6. T. von Kármán, *The Wind and Beyond* (1967 – ch. 3, n. 14), p. 271.
7. Alsop and Alsop, 'Push-Button War?' (n. 5).
8. D. I. Blumenstock, 'The Upper Atmosphere', *Scientific American*, vol. 180, January 1949; H. E. Roberts, 'The Earth's Atmosphere', *Aeronautical Engineering Rev.*, vol. 8, 1949. (In this book the word 'billion' is used to mean 'one thousand million'.)
9. W. R. Dornberger, *V-2* (New York: Viking, 1954), pp. xviii, 230.
10. DeHaven – Ltr, Maj. Gen. B. W. Chidlaw to Commanding General AAF (C. A. Spaatz), 12 May 1947, quoted in: E. Beard, *Developing the ICBM* (1976 – ch. 3, n. 44), p. 56.
11. A. C. Clarke, 'V-2 for Ionospheric Research?' *Wireless World*, February 1945; W. Ley, *Rockets* 2nd edn (1944 – ch. 3, n. 19), pp. 158–73.
12. C. P. Lent, *Rocket Research: History and Handbook* 2nd edn (New York: Pen-Ink Co., 1945), pp. 77–9.
13. P. Wright, *Spycatcher* (New York: Viking Penguin, 1987), pp. 11–12.
14. von Braun, 'Survey' (n. 3); US Department of Defense, H. H. Arnold, *Third Report to the Secretary of War* (1945 – ch. 3, n. 39), p. 68; H. Baldwin, 'Orbit Rockets Forecast', *New York Times*, 21 August 1946.
15. R. C. Hall, 'Earth Satellites, A First Look by the United States Navy' in R. C. Hall (ed.), *History of Rocketry and Astronautics* (1977 – ch. 3, n. 51), p. 259.
16. D. A. Rosenberg, 'The Origins of Overkill', *International Security*, vol. 7, no. 4, 1983, p. 15.
17. J. A. O'Keefe, 'Geodesy Comes of Age with Vanguard', *Astronautics*, vol. 2, no. 1, August 1957, p. 143; B. Price, 'Our Maps Err; Our Missiles Must Miss', *Washington Post*, 27 April 1958.
18. US Congress, Senate, Armed Services, Hearings: *Inquiry into Satellite and Missile Programs* (1958 – ch. 1, n. 11), testimony of James Kindelberger, chairman of North American Aviation, p. 1285.
19. US Department of Defense, H. E. Landsberg, *Geophysics and Warfare*, report for the Research and Development Coordinating Committee on General Sciences, CGS 202/1, 1954, pp. 10–13, 26–7. A search of the US National Archives failed to trace a copy of the earlier version: *Geophysical Sciences and Warfare*, report for the Geophysical Sciences Committee of the Research and Development Board, RDB 142/1, 1948.
20. See for example testimony from National Science Foundation officers on 1 July 1954: US Congress, House, Appropriations, Subcommittee on Independent Offices, Hearings: *Supplemental Appropriation Bill for 1955* pt 2 (84th Congress, 1st Session, 1955), pp. 895 ff.
21. NHO – Berkner Biog. File, L. V. Berkner, 'Science and the World Tomorrow', mimeo of lecture to Industrial College of the Armed Forces, 19 January 1962. The late James Killian recalled that Berkner was advocating the design of satellites for detecting nuclear explosions 'in the early '50s': interview.
22. Apart, that is, from Berkner's own similar but vaguer comment made a

few days earlier in: L. V. Berkner, *Outer Space: Prospects for Man and Society*, address at Seventh Southern Assembly, 11–14 January 1962 (Tulane, LA: Southern Assembly, 1962).

23.  The account of early BuAer satellite studies given here is largely based on R. C. Hall, 'Early U.S. Satellite Proposals', *Technology & Society*, vol. 4, no. 4, 1963; see also revised version in E. M. Emme (ed.), *The History of Rocket Technology: Essays on Research, Development, and Utility* (Detroit: Wayne State U. Press, 1964), and Hall, 'Earth Satellites' (n. 15). Hall's work was based on correspondence with Cogswell and Haviland, and on an unpublished historical MS by Captain (later Rear Admiral) Delmer Fahrney, head of BuAer's Special Design Branch for guided missiles, which the present author has not been able to consult: D. S. Fahrney, 'The History of Pilotless Aircraft and Guided Missiles', unpublished (BuAer, 1958).

24.  Memo. R. Haviland for J. A. Chambers, 'Comments Concerning Use and Development of Rockets', Aer-E-313-RPH, F.41 (1), 10 August 1945, cited in: Hall, 'Earth Satellites' (n. 15), p. 254.

25.  Hall, 'Earth Satellites' (n. 15), pp. 268–9. The first published suggestion that 'artificial satellite[s]' in what are now known as geosynchronous orbits might provide 'the *ultimate* solution to the problem' of global communications 'perhaps half a century ahead' was made by Flight Lieutenant Arthur C. Clarke (RAFVR) in a letter in *Wireless World*: Clarke, 'V-2?' (n. 11) – emphasis in original. Haviland later recalled having seen either this piece or Clarke's subsequent article in the same journal: A. C. Clarke, 'Extra-Terrestrial Relays', *Wireless World*, October 1945. Like many others, Clarke was thinking about manned communications satellites, and indeed continued to do so for over twenty years: A. C. Clarke, *The Promise of Space* (New York: Harper & Row, 1968), p. 100.

26.  G. H. Osborn, R. Gordon and H. L. Coplen, with G. S. James, 'Liquid-Hydrogen Rocket Engine Development at Aerojet, 1944–1950' in Hall (ed.), *Rocketry and Astronautics* (n. 15).

27.  R. G. Albion and R. H. Connery, *Forrestal and the Navy* (New York: Columbia U. Press, 1962), p. 154. Estimates for the Navy project were between $5 and $8 million.

28.  US AAF Air Technical Service Command, Engineering Division, 'Proposed Air Engineering Development Center', 10 December 1945, quoted in: Hall, 'Earth Satellites' (n. 15), p. 273.

29.  Project RAND, Clauser (ed.), *World-Circling Spaceship* (n. 4).

30.  Hall, 'Earth Satellites' (n. 15), p. 259.

31.  NHO – R. M. Salter, OH Interview with B. Ramsell, May 1961, pp. 24, 28.

32.  M. E. Davies and W. R. Harris, *RAND's Role in the Evolution of Balloon and Satellite Observation Systems and Related U.S. Space Technology* (Santa Monica: RAND, 1988), pp. 18–27; RAND Corporation, Social Sciences Division, *Proc. Conference on Methods for*

*Studying the Psychological Effects of Unconventional Weapons* (26–8 January 1949), Ref. D-387, 3 February 1949; RAND Corporation, P. Kecskemeti, *The Satellite Rocket Vehicle: Political and Psychological Problems*, Ref. RM-567, 4 October 1950.

33. Davies and Harris, *RAND's Role* (n. 32), p. 44.
34. E. M. Rogers, 'Man-Made Satellites', *Army Ordnance*, vol. 31, no. 159, November/December 1946.
35. Lang, 'What's Up There?' (n. 1).
36. J. A. Van Allen, 'The Use of Rockets in Upper Atmospheric Research' in Transactions of the International Association of Terrestrial Magnetism and Electricity (Oslo, 27–28 August 1948), *IATME Bulletin*, no. 13, 1950.
37. SAOHP – J. A. Van Allen Oral History, Interviews with D. H. DeVorkin and A. Needell, April-August 1981, pp. 166–7; see also J. A. Van Allen, *Origins of Magnetospheric Physics* (1983 – ch. 1, n. 2), p. 33.
38. US Department of Defense, *First Report of the Secretary of Defense*, 29 December 1948, p. 129. For contemporary criticism of the 'Forrestal Announcement', see RAND, *Proc. Conference on Psychological Effects* (n. 32), pp. 80, 102–5.
39. *Business Week*, 8 January 1949.
40. 'Rocket to the Moon', *Life*, 17 January 1949.
41. W. Ley, 'Rockets', *Scientific American*, vol. 180, May 1949; W. Ley, 'The Satellite Rocket', *Technology Review*, vol. 52, no. 2, December 1949; W. Ley with Chesley Bonestell, *The Conquest of Space* (New York: Viking, 1949).
42. K. W. Gatland, 'The Research Scene', *J. Brit. Interplanetary Soc.*, vol. 8, no. 4, July 1949.
43. K. W. Gatland, A. E. Dixon and A. M. Kunesch, 'Initial Objectives in Astronautics', *J. Brit. Interplanetary Soc.*, vol. 9, no. 4, July 1950; K. W. Gatland, A. M. Kunesch and A. E. Dixon, 'Minimum Satellite Vehicles', *J. Brit. Interplanetary Soc.*, vol. 10, no. 6, November 1951; repr. L. J. Carter (ed.), *The Artificial Satellite*, Proc. 2nd IAF Congress, London, September 1951 (London: British Interplanetary Society, 1951). Fred Singer, an upper-atmosphere scientist who had worked with Van Allen at APL and was then on the staff of the ONR Information Office in London, was actually introduced to the idea of using unmanned instrumented satellites, which he later did much to popularize, by members of the Society: interview.
44. Some early satellite calculations were criticized a few years later by scientists from the Naval Research Laboratory's Viking research rocket programme: M. W. Rosen and R. B. Snodgrass, 'Margin for Error' in *Space-Flight Problems*, Proc. 4th IAF Congress, Zürich, August 1953 (Biel-Bienne: Laubscher et Cie, 1954); M. W. Rosen, 'A Down-To-Earth View of Space Flight', address to 2nd Hayden Planetarium Symposium on Space Flight, 13 October 1952, (November

1952 – revised version, mimeo in the author's collection); also revised, same title, *J. Brit. Interplanetary Soc.*, vol. 12, no. 1, January 1953.

45. H. E. Newell, 'A Review of Upper Atmosphere Research from Rockets', *Trans. Am. Geophysical Union*, vol. 31, no. 1, February 1950, p. 33.

46. A. C. Clarke, *Interplanetary Flight: an Introduction to Astronautics* (London: Temple Press, 1950).

47. *J. Brit. Interplanetary Soc.*, vol. 11, no. 1, January 1952, p. 4.

48. W. von Braun, 'Crossing the Last Frontier', *Collier's*, 22 March 1952; W. von Braun, 'Prelude to Space Travel' in C. Ryan (ed.), *Across the Space Frontier* (London: Sidgwick and Jackson, 1952).

49. C. G. Lasby, *Project Paperclip: German Scientists and the Cold War* (New York: Atheneum, 1971), pp. 185–207.

50. W. von Braun, 'Multi-Stage Rockets and Artificial Satellites,' paper to U. Illinois Symposium on Space Medicine, March 1950, in J. P. Marbarger (ed.), *Space Medicine* (Chicago: U. Illinois Press, 1951); W. von Braun, 'The Importance of Satellite Vehicles in Interplanetary Flight', paper to 2nd IAF Congress, London, September 1951, *J. Brit. Interplanetary Soc.*, vol. 10, no. 6 (supplement – paginated as for no. 5), November 1951.

51. F. I. Ordway and M. R. Sharpe, *The Rocket Team* (1979 – ch. 3, n. 38), p. 361. Von Braun's emergence into the limelight in 1951–2 was also connected to the Army's decision in 1949 to move the German rocket experts from Fort Bliss, El Paso, to the Redstone Arsenal at Huntsville, there to pursue a serious development programme for tactical surface-to-surface ballistic missiles.

52. W. von Braun, 'Let's Tackle the Space Ship', extracts from lecture at Naval Ordnance Laboratory, 19 March 1952, *J. Am. Rocket Soc.*, vol. 22, no. 3, May/June 1952.

53. W. von Braun, 'Man on the Moon – The Journey', *Collier's*, 18 October 1952; revised version in C. Ryan (ed.), *Man on the Moon* (London: Sidgwick and Jackson, 1953).

54. W. von Braun, 'Von Braun Offers Plan for Station in Space', address to 2nd Hayden Planetarium Symposium on Space Flight, 13 October 1952, *Aviation Age*, vol. 18, no. 6, December 1952.

55. W. von Braun, 'The Baby Space Station', *Collier's*, 27 June 1953 – emphases added.

56. Rosen, 'Down-To-Earth View' (n. 44); M. W. Rosen, 'Review of *Across the Space Frontier*' (1952 – mimeo in the author's collection).

57. *Time*, 8 December 1952.

58. W. von Braun, 'Space Superiority', *Ordnance*, vol. 37, no. 197, March/April 1953.

59. Rosen, 'Down-To-Earth View' (n. 44).

60. M. Caidin, *Worlds in Space* (London: Sidgwick and Jackson, 1954), pp. 139–41.

61. O. H. Gauer and H. Haber, 'Man under Gravity-free Conditions' in

*German Aviation Medicine of World War II* pt 1 (USAF School of Aviation Medicine, 1950); H. Strughold, 'From Aviation Medicine to Space Medicine' in K. F. Gantz (ed.), *Man In Space* (London: Hollis and Carter, 1959).

62.    H. Armstrong, H. Haber and H. Strughold, 'Aeromedical Problems of Space Travel', *Aviation Medicine*, vol. 20, 1949; S. Thomas, *Men of Space* vol. 4 (Philadelphia: Chilton Books, 1962), p. 251.

63.    von Braun, 'Multi-stage Rockets' (n. 50), p. 26.

64.    C. S. White and O. O. Benson (eds), *Physics and Medicine of the Upper Atmosphere* (Albuquerque, NM: U. New Mexico Press, 1952), p. 480.

## Chapter 5: Intelligence for Truman

1.    D. Holloway, *The Soviet Union and the Arms Race* (London: Yale U. Press, 1983), pp. 21–2.

2.    J. McGovern, *Crossbow and Overcast* (London: Hutchinson, 1964); F. I. Ordway and M. R. Sharpe, *The Rocket Team* (1979 – ch. 3, n. 38), pp. 318–43.

3.    Ordway and Sharpe, *Rocket Team* (n. 2), ibid.; B. Harvey, *Race Into Space* (1988 – ch. 1, n. 32), pp. 21–3.

4.    G. A. Tokaev, *Stalin Means War* (London: Gutenberg Press, 1951), pp. 100–5; Holloway, *Soviet Union* (n. 1), pp. 23, 150–1.

5.    Holloway, *Soviet Union* (n. 1), p. 151.

6.    B. Gunston, *The Illustrated Encyclopædia of the World's Rockets & Missiles* (London: Salamander Books, 1979), p. 26.

7.    M. Stoiko, *Soviet Rocketry* (New York: Holt, Rinehart and Winston, 1970), pp. 74–5; Harvey, *Race* (n. 3), p. 23; Gunston, *Rockets & Missiles* (n. 6), p. 50.

8.    D. Campbell, *The Unsinkable Aircraft Carrier: American Military Power in Britain* (London: Michael Joseph, 1984), pp. 152–3; J. T. Richelson and D. Ball, *The Ties That Bind: Intelligence Cooperation between the UKUSA Countries* (London: Allen & Unwin, 1985), pp. 4–5.

9.    M. R. Beschloss, *Mayday: Eisenhower, Khrushchev and the U-2 Affair* (New York: Harper & Row, 1986), pp. 77–8.

10.    Quoted from the Soviet *Military Herald* in R. L. Garthoff, 'Russia – Leading the World in ICBM and Satellite Development', *Missiles and Rockets*, vol. 2, no. 10, October 1957.

11.    W. Ley, *Rockets, Missiles, and Space Travel* (New York: Viking, 1957), pp. 11–12.

12.    A. Parry, *Russia's Rockets and Missiles* (London: Macmillan, 1960), p. 70.

13.    US Strategic Bombing Survey, *Summary Report (Pacific War)*, 1 July 1946, p. 31.

14. US Congress, House, Appropriations, Hearings: *Military Appropriation for 1948* (1947 – ch. 2, n. 9), p. 649. LeMay was referring to sightings of 'strange, cigar-shaped objects . . . [by] Witnesses in Sweden and Finland . . . close to the Soviet border . . . ': D. M. Jacobs, *The UFO Controversy in America* (New York: New American Library, 1976), p. 31; c.f. *New York Times*, 12–14 August 1946.

15. HSTL – Papers of President's Air Policy Commission, Box 41, Folder: MH3–4, Ltr, Gallery to McCone, 22 October 1947.

16. HSTL – ibid., Memo. no. 39, McCone for Boatner and Pihl; HSTL – ibid., report by Boatner and Pihl, 'Data for the President's Air Policy Commission Concerning Guided Missiles', 28 October 1947 (signed by Maj. Gen. E. E. Partridge, Acting Deputy Chief of Staff – Operations).

17. USNA – RG 165, Records of the War Department General and Special Staffs, ABC 471–6 (7 Oct. 1943), (Sec 1–E) October 1947, quoted in: T. Bower, *The Paperclip Conspiracy* (1987 – ch. 2, n. 5), p. 271.

18. USNA – RG 319, Army Intelligence Decimal File 1941–48, 400.112 Research (032) 21.10.48., US Army Intelligence Report, 21 October 1948, quoted in: Bower, *Conspiracy* (n. 17), p. 271.

19. HSTL – President's Secretary's Files, Intelligence, Box 258, Folder: O.S.I./S.R., CIA Office of Scientific Intelligence, 'Soviet Flame and Combustion Research and its Relation to Jet Propulsion (Including Rocket Propulsion)', OSI/SR-6/49, 10 November 1949.

20. Tokaev, *Stalin* (n. 4), pp. 100, 105; see also G. A. Tokaev, 'Stalin demands a "round the world" rocket', *Sunday Express*, 23 January 1949.

21. E. Sänger and I. Bredt, *Über Einen Raketenantrieb für Fernbomber* (Ainbring: Deutsche Luftfahrtforschung, 1944).

22. Tokaev, *Stalin* (n. 4), p. 104.

23. A former colleague of Tokaev at the City College London has told the author that Tokaev was markedly prone to errors of nomenclature, both in the case of the V-weapons and with other rocket types: personal communication.

24. G. A. Tokaev, *Comrade X* (London: Harvill Press, 1956); G. A. Tokaty [i.e. Tokaev], 'Soviet Space Technology', *Spaceflight*, vol. 5, no. 2, March 1963. (Tokaev changed his surname to 'Tokaty' at some point between these two dates, perhaps in order to revert from a Russified to an original Ossetian form. American space historians have been excessively confused by this, inventing the spurious double-barrelled name 'Tokaty-Tokaev' and even the variant 'Tokady', neither of which Tokaev ever signed. Apart from references to work signed with his later name, he is referred to as 'Tokaev' throughout.)

25. Edmund Beard's suggestion that Tokaev's information only penetrated American intelligence circles *after* the joint intelligence assessments made in 1951–2 is highly implausible: *Developing the ICBM* (1976 – ch. 3, n. 44), p. 6.

26. *New York Times*, 2 May 1950.

27.    E. G. Schwiebert, *A History of the U.S. Air Force Ballistic Missiles* (New York: Praeger, 1965), pp. 57–60.
28.    DD-1976 – 246A: report by Joint Strategic Plans Committee to the Joint Chiefs of Staff, 'Re-examination of Programs for National Security', 25 October 1952 – emphasis added.
29.    DD-1975 – 5I: US Central Intelligence Agency, 'A Summary of Soviet Guided Missile Intelligence', US/UK-GM-4-52, 20 July 1953.
30.    Ltr, General D. L. Putt to Director of R&D, USAF Headquarters, 'Long Range Rocket Pilotless Aircraft Program, Atlas (MX-1593)', 20 March 1952, quoted in: Beard, *ICBM* (n. 25), pp. 137–8.
31.    DeHaven – Memo. T. Gardner for Assistant Secretary of Defense (R&D), 16 February 1954, quoted in: Beard, *ICBM* (n. 25), p. 164.
32.    Schwiebert, *Ballistic Missiles* (n. 27), p. 67; Beard, *ICBM* (n. 25), p. 6.
33.    R. L. Perry, 'The Atlas, Thor, Titan and Minuteman' in E. M. Emme (ed.), *The History of Rocket Technology* (1964 – ch. 4, n. 23), pp. 142–3; W. von Braun and F. I. Ordway, *History of Rocketry & Space Travel* (1967 – ch. 1, n. 45), p. 179.
34.    E. M. Emme, *A History of Space Flight* (New York: Holt, Rinehart and Winston, 1965), p. 108.
35.    US Congress, House, Government Operations, Hearings: *Organization and Administration of the Military R&D Programs* (83rd Congress, 2nd Session, 1954), p. 464.
36.    D. Masters and K. Way (eds), *One World Or None* (1946 – ch. 3, n. 28), pp. 44, 54.
37.    Jacobs, *Controversy* (n. 14), pp. 36–43. There are numerous tantalizing links between the history of UFO sightings and the history of early American space policy which cannot be explored here. For example, both Joseph Kaplan and Lloyd Berkner, two prominent members of the US IGY Committee, had acted as advisers to the USAF's Project Blue Book on UFOs in the early 1950s before they began to lobby openly for an IGY satellite project.
38.    D. Pearson and J. Anderson, *USA – Second-Class Power?* (1958 – ch. 1, n. 40), p. 66.
39.    W. A. McDougall, . . . *the Heavens and the Earth* (1985 – ch. 1, n. 47), p. 97.
40.    W. von Braun, F. I. Ordway and D. Dooling, *Space Travel* (1985 – ch. 1, n. 46), p. 170.

## Chapter 6: The Truman Space Policy

1.    Where specific references have not been given, the author acknowledges the following secondary sources for this chapter: E. Beard, *Developing the ICBM* (1976 – ch. 3, n. 44); J. L. Chapman, *Atlas* (1960 – ch. 3, n. 49); E. M. Emme, *Aeronautics and Astronautics: an*

*American Chronology of Science and Technology in the Exploration of Space 1915–1960* (Washington: NASA, 1961); S. L. Rearden, *The Formative Years* (1984 – ch. 2, n. 22).

2.  US Department of Defense, *First Report of the Secretary of Defense*, 29 December 1948, p. 9.

3.  US Department of Defense, H. H. Arnold, *Second Report of the Commanding General of the U.S. Army Air Forces to the Secretary of War* (London: HMSO, 1945), p. 94. James Doolittle had similar ideas to Arnold's, and the necessary status as a legendary aviator to promote them, but he also left the service in 1946.

4.  D. Mackenzie and G. Spinardi, 'The Shaping of Nuclear Weapon System Technology: US Fleet Ballistic Missile Guidance and Navigation – 1: From Polaris to Poseidon', *Social Studies of Science*, vol. 18, no. 3, 1988.

5.  Marginal notes by Maj. Gen. B. W. Chidlaw, quoted in: Beard, *ICBM* (n. 1), p. 111.

6.  DDEL – Harlow Papers, Box 2, Folder: Missiles, Misc. 1957–58 (1), Ltr, Truman to Patterson, 1 August 1946, in: 'Fiscal Year 1947 Funds Impounded by Truman,' documents circulated by the Senate Republican Policy Committee, February 1958.

7.  DeHaven – Ltr, Maj. Gen. B. W. Chidlaw to Commanding General AAF (C. A. Spaatz), 12 May 1947, quoted in: Beard, *ICBM* (n. 1), p. 54.

8.  Memo. Brig. Gen. T. S. Power for Commanding General AAF (C. A. Spaatz), 12 June 1947, quoted in: Beard, *ICBM* (n. 1), pp. 57–61. The order of missile priorities thus established was later endorsed by civilian experts in the Technical Evaluation Group attached to the RDB's Guided Missiles Committee: Beard, *ICBM* (n. 1), pp. 84–5.

9.  Beard, *ICBM* (n. 1), p. 67.

10. US President's Air Policy Commission, Report: *Survival in the Air Age*, 1 January 1948, pp. 82–3.

11. Beard, *ICBM* (n. 1), p. 75.

12. Rearden, *Formative Years* (n. 1), p. 102.

13. USAF Air Materiel Command Archives, File K 201–82, vol. 2, December 1954, report of a Special Committee of the Scientific Advisory Board to the Chief of Staff (USAF), 'Research and Development in the USAF', 21 September 1949, quoted in: Beard, *ICBM* (n. 1), p. 110.

14. USAF Air University Study, 'Research and Development in the United States Air Force', 18 November 1949, quoted in: Beard, *ICBM* (n. 1), p. 114.

15. Memo. by Maj. Gen. D. L. Putt, 23 August 1949, quoted in: Beard, *ICBM* (n. 1), p. 100.

16. Beard credits Lyndon Johnson with the leading role in this agitation. This view places too much weight on an isolated speech made by Johnson in February 1950 (*New York Times*, 14 February 1950), in which the freshman senator said that the Russian missile programme

was two or more years ahead of the American: Beard, *ICBM* (n. 1), p. 122.

17. Articles by Arthur Krock, *New York Times*, 1 and 5 November 1957.
18. Ibid. (1 November).
19. HSTL – Keller Papers, Box 1, Folder: Guided Missile correspondence 1950–52, Ltr and 'Report to the President On Status of Guided Missiles', 10 July 1951. Neither Beard, *ICBM* (n. 1), nor W. A. McDougall, ... *the Heavens and the Earth* (1985 – ch. 1, n. 47) refers to this document.
20. HSTL – Presidential Papers, Name Files, Box 45, Ltr, Keller to Eisenhower, 12 May 1953.
21. F. I. Ordway and M. R. Sharpe, *The Rocket Team* (1979 – ch. 3, n. 38), pp. 371–2.
22. US Congress, Senate, Armed Services, Hearings: *Inquiry into Satellite and Missile Programs* (1958 – ch. 1, n. 11), pp. 584–5.
23. Chapman, *Atlas* (n. 1), p. 63 – emphases added.
24. US Congress, House, Astronautics, Hearings: *Astronautics and Space Exploration* (85th Congress, 2nd Session, 1958), pp. 1504 ff.
25. HSTL – Post-Presidential Papers, Secretary's Office File, Box 16, 'Final Report of the Director of Guided Missiles', 17 September 1953.
26. NHO – B. N. Harlow, OH Interview with E. M. Emme, 11 June 1974.
27. The overall account in this section is based on R. C. Hall, 'Early U.S. Satellite Proposals' (1963 – ch. 4, n. 23), and 'Earth Satellites, A First Look by the United States Navy' (1977 – ch. 4, n. 15).
28. TEG report, 'Satellite Vehicle Program', 29 March 1948, quoted in: Emme, *Astronautics* (n. 1), p. 59.
29. Emme, *Astronautics* (n. 1), p. 60.
30. US Department of Defense, *First Report* (n. 2), p. 129.
31. RAND Corporation, Social Sciences Division, *Proc. Conference on Psychological Effects of Unconventional Weapons* (1949 – ch. 4, n. 32), p. 102.
32. RAND Corporation, Social Sciences Division, *Summary Report on Conference on Psychological Effects of Unconventional Weapons* (26–8 January 1949) Ref. D-405, 21 February 1949, p. 14.
33. NHO – R. M. Salter, OH Interview with B. Ramsell, May 1961, p. 24.
34. NHO – H. E. Newell, OH Interview with E. M. Emme and A. F. Roland, 7 January 1974, pp. 7–8.
35. RAND Corporation, P. Kecskemeti, *The Satellite Rocket Vehicle* (1950 – ch. 4, n. 32).
36. Memo. USAF Deputy Chief of Staff (Materiel) for Vice Chief of Staff, 'Earth Satellite Vehicles', 8 January 1948, quoted in: McDougall, *Heavens* (n. 19), pp. 107–8.
37. Salter, OH Interview (n. 33), p. 22.

38. Memo. Chief, Guided Missile Section, Engineering Division, AMC, for Chief of Staff (USAF), 23 March 1949, quoted in: Beard, *ICBM* (n. 1), p. 92.

39. Project RAND, J. E. Lipp, 'The Time Factor in the Satellite Program', Ref. RAD-51, 18 October 1946, p. 1.

40. RAND Corporation, Social Sciences Division, *Proc. Conference* (n. 31), p. 12 – emphasis added.

41. RAND Corporation, Kecskemeti, *Satellite Rocket* (n. 35), pp. 21–2.

42. The surviving documents do not make it clear, but on reading between the lines it seems likely that the idea of this consultancy originated from Grosse himself. The wording of a subsequent letter from Grosse to the NASA historian Eugene Emme (NHO – Grosse Biog. File, 12 January 1973) strongly suggests that Grosse never actually met the President. Space historians have stated that Grosse was 'a key figure in the Manhattan Project in its early days': C. Mc. Green and M. Lomask, *Vanguard: a History* (Washington: Smithsonian Institution, 1971), p. 16. But Grosse never worked in any part of the Manhattan District. The story that he had done so arose from a letter written by von Braun to Robert Truax in 1952 which has not been checked as it should have been: S. Thomas, *Men of Space* vol. 2 (Philadelphia: Chilton Books, 1961), pp. 192–3.

43. HSTL – Miscellaneous Historical Documents, Box 3, A. V. Grosse, 'Report on the Present Status of the Satellite Problem', 25 August 1953, p. 1.

44. Copies also reached the ex-President, the White House, and Donald Quarles at the Department of Defense.

45. Grosse, 'Report' (n. 43), pp. 2, 3, 6, 7.

46. *New York Times*, 5 February 1956.

47. NHO – American Rocket Society Files, 'Proc. 1st Meeting of Ad Hoc Committee on Space Flight', 17 May 1952, p. 102.

48. Ibid., pp. 79–80.

49. H. S. Truman, *Year of Decisions* (London: Hodder and Stoughton, 1955); H. S. Truman, *Years of Trial and Hope* (London: Hodder and Stoughton, 1956).

50. H. S. Truman, *Mr Citizen* (London: Hutchinson, 1961).

51. NHO – Grosse Biog. File, Ltr, Graham to Grosse, 17 January 1973.

52. Hall, 'Early Satellite Proposals' (n. 27), p. 428; in his second essay on the early satellite studies Hall omitted any reference to the 'Forrestal announcement': Hall, 'Earth Satellites' (n. 27).

53. The usual reference is to E. Tarle, 'War Vs. Atomic Blackmail', *New Times*, November 1949.

54. For example: Chapman, *Atlas* (n. 1); Emme, *Astronautics* (n. 1); E. G. Schwiebert, *A History of the U.S. Air Force Ballistic Missiles* (1965 – ch. 5, n. 27); W. von Braun and F. I. Ordway, *History of Rocketry & Space Travel* (1967 – ch. 1, n. 45).

55. Emme, *Astronautics* (n. 1), p. 66; R. L. Perry, 'The Atlas, Thor, Titan

and Minuteman' (1964 – ch. 5, n. 33), pp. 142–3; C. C. Alexander, 'Part One – Research' in L. S. Swenson, J. M. Grimwood and C. C. Alexander, *This New Ocean: a History of Project Mercury* (Washington: NASA, 1966) p. 22; von Braun and Ordway, *History* (n. 54), p. 122.

56. Beard, *ICBM* (n. 1), p. 136.
57. McDougall, *Heavens* (n. 19), p. 105.
58. Thomas, *Men of Space* (n. 42), pp. 192–3.
59. Green and Lomask, *Vanguard* (n. 42), p. 16; McDougall, *Heavens* (n. 19), pp. 118–19.

## Chapter 7: Proposing Satellites for the IGY

1. *J. Geophysical Research*, vol. 54, no. 2, June 1949, p. 203.
2. US Department of State, L. V. Berkner, *Science and Foreign Relations* (1950 – ch. 3, n. 15), pp. 20, 21.
3. J. A. Fleming and V. Laursen, 'International Polar Year of 1932–33', *Science*, vol. 110, 23 September 1949.
4. W. Sullivan, *Assault on the Unknown: the International Geophysical Year* (London: Hodder and Stoughton, 1962), pp. 17–19.
5. M. Nicolet, 'The International Geophysical Year (1957–1958): Great Achievements and Minor Obstacles', *GeoJournal*, vol. 8, no. 4, 1984, pp. 308–9 – emphasis added.
6. D. H. DeVorkin, 'Organizing for space research: the V-2 rocket panel', *Hist. Stud. Phys. Biol. Sciences*, vol. 18, no. 1, 1987.
7. SAOHP – J. A. Van Allen Oral History, Interviews with D. H. DeVorkin and A. Needell, April–August 1981, p. 189.
8. H. E. Newell, 'Review of Upper Atmosphere Research' (1950 – ch. 4, n. 45), p. 33.
9. C. Mc. Green and M. Lomask, *Vanguard* (1971 – ch. 6, n. 42), p. 19.
10. In public Chapman always allowed Berkner to take the credit for the original IGY proposal, and even extended it to him in print: S. Chapman, *IGY: Year of Discovery* (Ann Arbor, MI: U. Michigan Press, 1959), p. 101. His private recollection, however, was simply that when he asked Berkner who had made the suggestion, 'he wrote back that it was he, so I took his word. I have heard that Van Allen said it was I, but I could certainly not say who it was.': Ltr, Chapman to T. G. Cowling, 14 April 1970, Chapman Papers, Bodleian Library, Oxford.
11. M. Nicolet, 'Historical Aspects of the IGY', *EOS*, vol. 64, no. 19, 10 May 1983.
12. Since the 1950s the structure of the international scientific organizations has been roughly three-tiered: an international 'association' is centred around a single research field or discipline; a group of formally or informally linked associations, or sometimes of its own internal sections or commissions, makes up a 'union'; and overall coordination,

including the occasional recognition of new unions, is effected by the International Council of Scientific Unions (ICSU). Thus in the early 1950s the International Association of Terrestrial Magnetism and Electricity (which was renamed the International Association of Geomagnetism and Aeronomy during the run-up to the IGY) was a member organization within the International Union of Geodesy and Geophysics, which in turn was one of about a dozen unions represented in ICSU: K. O. Murra, *International Scientific Organizations: a Guide to their Library, Documentation, and Information Services* (Washington: Library of Congress, 1962).

13. Nicolet, 'Historical Aspects' (n. 11) and 'Year' (n. 5); US Department of Defense, N. C. Gerson and R. J. Donaldson (eds), *Proc. Conference on Ionospheric Physics, Part A*; L. Katz and N. C. Gerson (eds), *Proc. Conference on Ionospheric Physics, Part B* (Cambridge, MA: Air Force Cambridge Research Center, 1952).

14. Nicolet, 'Year' (n. 5), p. 311.

15. L. V. Berkner, 'International Scientific Action: the International Geophysical Year 1957–58', *Science*, vol. 119, 30 April 1954.

16. The MCI's original draft programme is in *Annals of the IGY* (London: Pergamon, 1958–70 – hereafter *Annals*), vol. IIA (1959), pp. 69 ff., but wrongly dated to 1951. Likewise, at *Annals*, vol. I (1959), p. 383, the MCI meeting is wrongly dated to July instead of September 1950. It is appropriate to comment at this point that the *Annals* are frequently unreliable or inconsistent on minor points of organizational detail. Although the work of international scientific organizations for the IGY was often rather formal on the surface, there seems to have been considerable scope for repeated *ad hoc* adjustments to official decisions. This common-sense approach to the task at hand may have been necessary and even admirable at the time, but it greatly complicates the task of historical research into the IGY's decision-making processes.

17. This letter, now in the Royal Society's Council Minutes, vol. 18, appears to contradict the statement at *Annals*, vol. I (1959), p. 384, that initial invitations were not sent out until after the ICSU Executive had set up CSAGI on 16 May 1952. However, the Royal Society may have been favoured with an unusually early invitation, since Hill was also its Foreign Secretary at that time.

18. S. Chapman, 'The International Geophysical Year 1957–58', *Nature*, vol. 172, 22 August 1953; Chapman, *IGY* (n. 10), p. 101.

19. DDEL – WHO: NSC Staff, OCB Central Files, Box 11, OCB 000.91, Folder: 1 (2), Memo. J. W. Joyce for R. Hirsch, attachment, R. Hirsch, 'Staff Study on Coordination of Public Information on U.S. IGY Programs, 1957–58', n.d. but c. December 1955, p. 3.

20. There were really three CSAGIs. The elected IGY Committee rarely met as such, except to conduct formal business during the five roughly annual 'CSAGI meetings' held between 1953 and 1958. Most of the activity at these events consisted in meetings of *ad hoc* working groups

composed of delegates from the national IGY committees, who were not necessarily members of CSAGI as such. Thirdly, the Officers Bureau supervised the day-to-day work of the Secretary General and a small staff in Brussels.

21.  The posts of the Coordinator, Vice-Admiral Sir Archibald Day (RN, retd), and the Editor-in-Chief, former Astronomer Royal Sir Harold Spencer Jones, were financed by a special grant from the British Treasury: Royal Society, Council Minutes, 19 April 1956. The first of these appointments was related to the dissatisfaction with the management of CSAGI's affairs expressed by several delegations at the Barcelona meeting in 1956, with the British prominent amongst them. Members of the US Congress sometimes opined that the size of their country's IGY programme entitled it to a greater share of CSAGI membership: H. Bullis, 'The Political Legacy of the International Geophysical Year' in US Congress, House, International Relations, Committee Print: *Science, Technology, and American Diplomacy* (90th Congress, 1st Session, 1977), p. 307.

22.  P. Forman, 'Behind quantum electronics: national security as basis for physical research in the United States, 1940–1960', *Hist. Stud. Phys. Biol. Sciences*, vol. 18, no. 1, 1987, p. 211.

23.  J. A. Van Allen, 'Rockets in Upper Atmospheric Research' (1948 – ch. 4, n. 36); see also *Science News Letter*, 4 September 1948. The penultimate sentence, omitted from the published text, read as follows: 'Serious consideration is being given to the development of a satellite missile which will continuously orbit around the earth at a distance of, say, 1000 km': J. A. Van Allen, *Origins of Magnetospheric Physics* (1983 – ch. 1, n. 2), p. 33.

24.  Singer's first public lecture on the scientific satellite idea was given to the British Interplanetary Society on 10 November 1951, two months after Kenneth Gatland and his colleagues had presented their historic paper on 'minimum satellites' to the Second IAF Congress: S. F. Singer, 'Research in the Upper Atmosphere with Sounding Rockets and Earth Satellite Vehicles', *J. Brit. Interplanetary Soc.*, vol. 11, no. 2, March 1952; K. W. Gatland et al., 'Minimum Satellite Vehicles' (1951 – ch. 4, n. 43). Conferences which received presentations or papers from Singer included a British Association meeting in Belfast in September 1952, the Oxford conference in August 1953 (n. 26), and the Fourth IAF Congress in Zürich in September 1953.

25.  Nicolet, 'Year' (n. 5), p. 316.

26.  Berkner also attended the Oxford conference, though his name was omitted from the Proceedings volume: R. L. F. Boyd and M. J. Seaton (eds), *Rocket Exploration of the Upper Atmosphere* (1954 – ch. 3, n. 13), pp. 375–6. A copy of the programme used by Charles Green, a rocket engineer from General Electric (NHO – C. F. Green Biog. File), is annotated with Berkner's name as one of the contributors to the closing discussion on rocket and satellite experiments. As Singer

gave no technical details either to the British Interplanetary Society or in his Oxford paper, it is interesting to note that, according to Green's jottings, the notional satellite discussed at Oxford would have used a '16 000 kg 1st stage' with '30 000 kg thrust' to place a mass of 50 kg into orbit; no postulated altitude is mentioned.

27. M. E. Davies and W. R. Harris, *RAND's Role in the Evolution of Balloon and Satellite Observation Systems* (1988 – ch. 4, n. 32), pp. 53–5.

28. J. R. Killian, *Sputnik, Scientists, and Eisenhower* (1977 – ch. 3, n. 1), pp. 68, 69.

29. US National Security Council, 'U.S. Scientific Satellite Program', NSC 5520, drafted 20 May, approved 26 May 1955, in P. Kesaris (ed.), *Documents of the National Security Council* (microfilm) Reel II Item 0174 (Frederick, MD: University Publications of America, 1983); US National Security Council, report of the ODM-SAC Technological Capabilities Panel, 'Meeting the Threat of Surprise Attack', 14 February 1955, in P. Kesaris (ed.), *Minutes of Meetings of the National Security Council, with Special Advisory Reports* (microfilm) Reel II Items 0755 and 0805 (Frederick, MD: University Publications of America, 1983).

30. Singer recalls Berkner as having 'controlled' the URSI meeting: interview.

31. J. Holmes, *America on the Moon: the Enterprise of the Sixties* (Philadelphia: Lippincott, 1962), p. 48.

32. According to Singer, 'satellites were a very non-respectable type of activity, thanks to von Braun's premature emphasis on manned space stations': interview. Van Allen recalls that the principal officers of the IGY would have viewed a satellite proposal as 'excessively speculative' at that point: interview.

33. NHO – Project Files, Folder: SCORE, H. K. Ziegler, 'Prelude to SCORE', pt 1 in H. K. Ziegler and H. McD. Brown, 'The Signal Corps' Pioneering Achievements at the Space Frontier' (mimeo – 1981), p. 7 – emphases added.

34. Green and Lomask, *Vanguard* (n. 9), p. 19.

35. The Academia Sinica of Beijing joined the IGY in late 1955, after the Soviet Academy, and took an active part in preparations over the next eighteen months. In June 1957, on the eve the Year's opening, the Beijing Academy withdrew because the Academia Sinica of Taipei had joined at the last minute, and Chapman's efforts to find a form of words which would enable both academies to take part were unsuccessful.

36. Nicolet, 'Year' (n. 5), Figs 17, 18.

37. *IUGG Newsletter*, 4th Year, no. 9, March 1955.

38. Green and Lomask, *Vanguard* (n. 9), p. 21. However, the Soviet adhesion, followed by the ill-starred membership of Communist China, prompted the addition of another meridian of longitude along which to concentrate observations, at 110° East.

39. An official version of Sedov's remarks was issued in order to correct

what were claimed to be inaccurate reports in the Western press, which represented him as declaring that there would be a Soviet satellite programme *for the IGY*. Translations of this and other relevant Soviet documents and a careful chronology of all Soviet statements were published in: F. J. Krieger, *Behind the Sputniks: a Survey of Soviet Space Science* (Washington: Public Affairs Press, 1958). Although the official Soviet announcement of *an IGY satellite project* was not made until September 1956 at Barcelona, several historians, neglecting Krieger, have given currency to the story that one was made at the end of July 1955; see for example: Bullis, 'Political Legacy' (n. 21), p. 333; W. A. McDougall, . . . *the Heavens and the Earth* (1985 – ch. 1, n. 47), p. 60, and B. Harvey, *Race Into Space* (1988 – ch. 1, n. 32), p. 27.

40. A. G. Karpenko and G. A. Skuridin (trans. A. Pingell), 'Contemporary Problems of Space Flying', *Vestnik Akademii Nauk*, September 1955 (Washington: Naval Research Laboratory, 1957).

41. *Annals*, vol. IIA (1959), pp. 208, 213–14. However, a confidential report on the 1955 CSAGI Working Group, written for the secretary of the US IGY Committee, Hugh Odishaw, stated that one Soviet scientist, N. Pinus, had at least attended as an observer, with an interpreter, though without contributing to the discussions: USNC-IGY – Earth Satellite Correspondence, Folder: Jan thru Sep 1955, Memo. P. H. Wyckoff for H. Odishaw, 'Report of Activity at CSAGI Meeting in Brussels', 26 September 1955.

42. *Annals*, vol. IIA (1959), p. 311.

43. *Proc. US Nat. Acad. Sci.*, vol. 40, 1954.

44. R. G. J. Fraser, *Once Round the Sun: the Story of the International Geophysical Year 1957–58* (London: Scientific Book Club, 1957); A. Marshack, *The World in Space: the Story of the International Geophysical Year* (New York: Nelson, 1958).

45. J. Kaplan, 'The IGY Program', *Proc. IRE*, vol. 44 (1), no. 6, 1956, p. 743; J. Kaplan and H. Odishaw, 'Satellite Program', *Science*, vol. 122, 25 November 1955, p. 1003; L. V. Berkner and H. Odishaw, 'Dimensions and Problems' in L. V. Berkner and H. Odishaw (eds), *Science in Space* (New York: McGraw-Hill, 1961), p. 36; J. P. Hagen, 'The Viking and the Vanguard' in E. M. Emme (ed.), *The History of Rocket Technology* (1964 – ch. 4, n. 23), p. 125.

46. See for example: E. M. Emme, *History of Space Flight* (1965 – ch. 5, n. 34), p. 116; E. C. B. Schoettle, 'The Establishment of NASA' in S. A. Lakoff (ed.), *Knowledge and Power: Essays on Science and Government* (New York: Free Press, 1966), p. 167; W. von Braun and F. I. Ordway, *History of Rocketry & Space Travel* (1967 – ch. 1, n. 45), pp. 179–80; C. C. Alexander, 'Part One - Research' (1966 – ch. 6, n. 55), pp. 20, 28–9.

47. W. Sullivan, 'The International Geophysical Year', *International Conciliation*, January 1959; Sullivan, *Assault on the Unknown* (n. 4), p. 32. Sullivan's omission of Berkner from this part of his narrative is the more

interesting in that several passages in the book suggest that he depended heavily on Berkner for inside information about the functioning of CSAGI.

48. Holmes, *America on the Moon* (n. 31), pp. 46–9. Kaplan had described the meeting in rather general terms at US Congress, House, Astronautics, Hearings: *Astronautics and Space Exploration* (1958 – ch. 6, n. 24), pp. 1143–4.

49. Green and Lomask, *Vanguard* (n. 9), p. 21.

50. McDougall, *Heavens* (n. 39), pp. 59–60, 118.

51. See for example: USNC-IGY – National Committee, Folder: Office of Defense Mobilization, unsigned memo. 'Some Notes on the United States IGY Effort', n.d., and Memo. D. Z. Beckler for A. S. Flemming, 'International Geophysical Year', 13 April 1954, p. 3; also DDEL – WHO: NSC Staff, OCB Central Files, Box 11, OCB 000.91, Folder: 1 (2), Ltr, Waterman to Quarles, 13 May 1955, p. 1.

## Chapter 8: Satellites in the IGY

1. US Congress, House, Armed Services, Hearings: *Investigation of National Defense Missiles* (1958 – ch. 3, n. 57), p. 4304.

2. F. J. Krieger, *Behind the Sputniks* (1958 – ch. 7, n. 39), pp. 330, 114.

3. USNC-IGY – National Committee, Folder: Satellite Correspondence January 1956, 'A Summary of Some Recent Public Information on USSR Rocket and Satellite Developments', January 1956.

4. DDEL – WHO: Secretary, White House – Subjects, Box 4, Folder: Dr Killian (1) May '56 – Nov. '57, Brig. Gen. A. J. Goodpaster, Memo. for Record, 16 May 1956.

5. USNC-IGY – TPESP, Folder: Satellite Correspondence July 1956, Ltr, H. Odishaw to M. Nicolet, 15 June 1956.

6. Ibid., Memo. to Files from J. Kaplan, 25 July 1956.

7. *New York Times*, 2 June 1957, reporting *Pravda* of previous day; *Komsomolskaya Pravda*, 9 June 1957, trans. Krieger, *Sputniks* (n. 2), pp. 10, 276–81 – emphasis added. In July Kenneth Gatland predicted a possible Soviet satellite launch attempt for 5 September: 'Rockets and Artificial Satellites of the I.G.Y.' *Spaceflight*, vol. 1, no. 4, July 1957.

8. P. J. Klass, *Secret Sentries in Space* (New York: Random House, 1971), pp. 30–1; M. R. Beschloss, *May-Day* (1986 – ch. 5, n. 9), p. 155; P. Pringle and W. Arkin, *SIOP: Nuclear War from the Inside* (London: Sphere Books, 1983), pp. 51–2.

9. USNC-IGY – TPESP, Folder: Satellite Correspondence June 1957, J. G. Reid, Memo. for Record, 4 June 1957.

10. Copies of the Soviet report in the original and in a re-typed version, with Bardin's covering letter of 10 June and Berkner's letter transmitting the report to Nicolet on 16 June, are at: USNC-IGY – TPESP, Folder:

Satellite Correspondence June 1957. The account given here rejects, as inconsistent with the documentary evidence, both the statement by von Braun and Ordway that Berkner *received* the Soviet report on 10 June: W. von Braun and F. I. Ordway, *History of Rocketry & Space Travel* (1967 – ch. 1, n. 45), p. 180, and the statement by NASA historians that Bardin handed the report to Berkner in Washington at the NAS conference held between 26 and 28 June to mark the opening of the IGY: C. Mc. Green and M. Lomask, *Vanguard* (1971 – ch. 6, n. 42), p. 129. The vague information in the report that Soviet satellites would be launched 'at a small angle to the meridian' and could be 'observed in all the areas of the Earth' was circulated in July: *IGY News*, 15 July 1957, Items 43 and 44.

11. USNC-IGY – TPESP, Folder: Satellite Correspondence June 1957, R. Porter, notes on a copy of the covering letter attached to the Soviet report (J. G. Reid to TPESP members, 24 June 1957) – emphasis in original.

12. US Congress, House, Astronautics, Hearings: *Astronautics and Space Exploration* (1958 – ch. 6, n. 24), p. 345.

13. Each World Data Centre was made up of separate centres for the IGY disciplines, based at various locations or institutions, and the third was dispersed between several countries. World Data Centre C for Rockets and Satellites was established at the British government's Radio Research Station at Slough in October 1958, after a successful application for this role had been made at the Moscow CSAGI meeting.

14. W. Sullivan, 'The International Geophysical Year' (1959 – ch. 7, n. 47), p. 307.

15. USNC-IGY – Reid, Memo. for Record (n. 9).

16. NAS – International Relations 1957, Folder: IGY General, Ltr, L. V. Berkner to R. Fraser, 29 July 1957.

17. The account given at this point is based largely on Sullivan, 'Year' (n. 14), pp. 308–12.

18. *Annals*, vol. VI, L. V. Berkner (ed.), *Manual on Rockets and Satellites* (1958), pp. 465, 468, 470.

19. *Annals*, vol. VII (1959), pp. 314–15; Sullivan, 'Year' (n. 14), p. 316.

20. For example: J. A. Van Allen (ed.), *Scientific Uses of Earth Satellites* (Ann Arbor, MI: U. Michigan Press, 1956); *Proc. IRE*, vol. 44, no. 6, 1956; *J. Franklin Institute*, Monograph no. 2, June 1956.

21. *Annals*, vol. VI (1958), p. 457.

22. H. S. W. Massey and M. O. Robins, *History of British Space Science* (Cambridge: Cambridge U. Press, 1986), p. 43; I. Southall, *Woomera* (London: Angus and Robertson, 1962), pp. 214–15.

23. The Czechs were at first obliged to follow the instructions published by the US IGY Committee for amateur 'Moonwatch' teams. Then the June *Radio* article published in the Soviet Union was translated in the October issue of the equivalent Czech journal *Amatérské*

*radio*: V. Guth, 'Rockets and Satellites' in Mathematico-Physical Section of Czechoslovak Academy of Sciences (eds), *International Geophysical Year and Cooperation in Czechoslovakia 1957–1959* (Prague: Nakladatelství Československé Akademie Véd, 1960).

24. Translations of the Soviet articles were published in *Annals*, vol. VI (1958), pp. 222–54. The copy of Bardin's letter to Berkner (16 August 1957) which went to the British IGY Committee (BCIGY – NGY/90 (57)) was in response to a cable from London a few days earlier pointing out that frequency details were 'a point of first importance': BCIGY – NGY/89 (57): Minutes of the Artificial Satellites Subcommittee, 8 August 1957, Item 4. Marcel Nicolet states that neither Bardin nor Berkner sent a copy to the CSAGI office in Brussels before Sputnik 1: interview.

25. BCIGY – NGY/117 (57): H. S. W. Massey, 'Report on the Washington Conference'.

26. *Annals*, vol. VI (1958), p. 467; c.f. item (1) (d) in the resolutions from the Washington Conference listed at ibid., p. 458.

27. In April 1958 Lloyd Berkner explained to the House Committee on Astronautics that restrictions on the exchange of information of this kind were mutually understood and accepted within the IGY context: US Congress, House, Astronautics, *Space Exploration* (n. 12), pp. 1036–7.

28. US Congress, Senate, Armed Services, Hearings: *Inquiry into Satellite and Missile Programs* (1958 – ch. 1, n. 11), p. 560.

29. *Annals*, vol. VI (1958), p. 467 – emphasis added.

30. *Annals*, vol. X (1960), p. 183.

31. Ibid., pp. 229–30; c.f. *Annals*, vol. VII (1959), p. 315.

32. Tadeo Takenouchi of Tokyo University subsequently tried to calculate the exact launching times and launch site of the first two sputniks from their orbital coordinates. The first official Soviet statement on the geographical coordinates of the Baikonur Cosmodrome (Tyuratam) was issued in May 1961: K. W. Gatland, *Astronautics in the Sixties* (1962 – ch. 1, n. 1), p. 139.

33. *Annals*, vol. VI (1958), p. 458.

34. Ibid., p. 467.

35. *Annals*, vol. X (1960), p. 183; Sullivan 'Year' (n. 14), p. 316.

36. *Annals*, vol. VI (1958), p. 470.

37. BCIGY – NGY/74 (56): Ltr, Hagen to Massey, 1 August 1956.

38. *Annals*, vol. VI (1958), p. 467.

39. Ibid., p. 460.

40. *Annals*, vol. VI (1958), pp. 464–5; for the Soviet response to these enquiries, see ibid., pp. 468–9.

41. *Annals*, vol. X (1960), pp. 181, 183; vol. VI (1958), p. 468.

42. *Annals*, vol. IX (1959), p. 231.

43. Ibid., p. 184.

44. *Annals*, vol. VI (1958), p. 470; vol. VII (1959), p. 317.

45. *Annals*, vol. VI (1958), p. 467 – emphasis added.
46. BCIGY – NGY/74 (56): Ltrs exchanged between Massey and Hagen.
47. US Congress, Senate, Space Sciences, Hearings: *NASA Authorization for Fiscal Year 1960* pt 1 (86th Congress, 1st Session, 1959), pp. 661–4; H. L. Richter, W. F. Sampson and R. Stevens, 'Microlock: a Minimum Weight Radio Instrumentation System for a Satellite' in M. Alperin and M. Stern (eds) *Vistas in Astronautics*, Proc. First Annual Air Force Office of Scientific Research Astronautics Symposium, San Diego, February 1957, (New York: Pergamon, 1958); W. H. Pickering with J. H. Wilson, 'Countdown to Space Exploration: a Memoir of the Jet Propulsion Laboratory' in R. C. Hall (ed.) *History of Rocketry and Astronautics* (1977 – ch. 3, n. 51). Both tracking systems were radio interferometers, but Minitrack, which rapidly developed into NASA's primary tracking and telemetry system, was significantly more accurate than Microlock, with a peak accuracy of 20–30 sec of arc by comparison with 3 to 4 min of arc: *Annals*, vol. VI (1958) pp. 279, 351, 409, 411.
48. Circular on US satellite programme from IGY Secretary General (Nicolet), 6 August 1956 (ICSU Archives, CN-CIR-15-568/6), p. 5; see also F. I. Ordway, 'The U.S. Satellite Vehicle Program', *Astronautica Acta*, vol. 2, no. 3, 1956.
49. *Annals*, vol. IIA (1959), p. 305.
50. W. Sullivan, *Assault on the Unknown* (1962 – ch. 7, n. 4), pp. 60–2.
51. RAND Corporation, H. Kallmann-Bijl and W. W. Kellogg, *Scientific Use of an Artificial Earth Satellite* (Santa Monica: RAND, 1955), p. 5.
52. See note 24 above.
53. *Annals*, vol. IIA (1959), p. 364.
54. Ibid., p. 343.
55. *Annals*, vol. X (1960), p. 181 – emphasis added.
56. Massey and Robins, *British Space Science* (n. 22), p. 39 – emphasis added
57. Adm. R. Bennett (USN), excerpts from address before the National Security Industrial Association, 8 October 1957, in US Congress, House, Government Operations, Hearings: *Availability of Information from Federal Departments and Agencies* (85th Congress, 1st Session, 1957), pp. 3203–4.
58. Sullivan, *Assault* (n. 50), p. 62.
59. BCIGY – NGY/82 (57): Ltr, L. V. Berkner to D. Martin, 19 July 1957.
60. *Annals*, vol. VI (1958), pp. 459, 461.
61. *Annals*, vol. X (1960), pp. 231–2.
62. From remarks made in Moscow by ICIC member Kirill Stanyukovitch on 5 October 1957: K. W. Gatland (ed.), *Project Satellite* (London: Allan Wingate, 1958), p. 102. Blagonravov also stated that Sputnik 1 was not an IGY satellite: *New York Times*, 7 October 1957.

63. DDEL – WHO: Staff Research Group, Box 15, Folder: N.S.F., draft memo. (probably by Waterman), 'Soviet Non-cooperation With IGY', 23 October 1957.

64. US Congress, House, Astronautics, *Space Exploration* (n. 12), pp. 824, 833.

65. Ibid., pp. 874–5. The Soviet IGY Committee's *Preliminary Report on Sputniks 1 and 2* was received at the CSAGI office on 3 February 1958: *IGY News*, 10 February 1958, Item 78. It reached the Washington Data Centre just over a week later: US Congress, Senate, Space Sciences, Report: *Soviet Space Programs: Organization, Plans, Goals and International Implications* (87th Congress, 2nd Session, 1962), p. 297.

66. US Congress, House, Astronautics, *Space Exploration* (n. 12), pp. 371–2.

67. Ibid., pp. 331–2, Ltr, H. Odishaw to J. P. Hagen, 9 May 1958.

68. DDEL – 'Soviet Non-cooperation With IGY' (n. 63).

69. *Annals*, vol. XII pt I (1960), pp. 15–16. The Japanese IGY Committee acknowledged that their members had made recordings of satellite telemetry, but only from three American Explorer satellites: *Japanese Contribution to the International Geophysical Year* (Tokyo: Science Council, 1958–61), vol. I p. 106; vol. II, p. 156; vol. III, p. 119.

70. S. N. Vernov, P. V. Vakulov, E. V. Gorchakov, Yu. I. Logachev and A. Ye. Chudakov, 'Study of Soft Component of Cosmic Rays beyond the Limits of the Atmosphere', *Annals*, vol. XII pt II (1961), p. 667.

71. BCIGY: Minutes of the Artificial Satellites Subcommittee, 15 October 1957, Item 3 (c) and Item 6; 14 May 1958, Item 4; Minutes of Radio Methods Working Group, 4 December 1958, Item 5.

72. NAS – Ltr, Berkner to Fraser (n. 16).

73. J. W. Dungey, 'The Radiation Belts' in D. R. Bates, *The Planet Earth* 2nd edn (Oxford: Pergamon Press, 1964), p. 311.

74. S. N. Vernov, N. L. Grigorov, Yu. I. Logachev and A. Ye. Chudakov, 'Measurement of cosmic radiation by rockets and artificial earth satellites', *Annals*, vol. XII pt II (1961), p. 652.

75. J. A. Van Allen: interview.

76. SAOHP – C. Y. Johnson Oral History, Interview with A. Needell and P. Shea, 21 April 1982, p. 27.

77. Sullivan, 'Year' (n. 14), pp. 211, 214.

78. US Congress, House, Astronautics, *Space Exploration* (n. 12), p. 354; Senate, Space Sciences, *Soviet Space Programs* (n. 65), p. 134.

79. US Congress, House, Astronautics, *Space Exploration* (n. 12), p. 1068; and Report: *International Cooperation in the Exploration of Space* (86th Congress, 1st Session, 1959), p. 15.

80. 'Catalogue of Data in the World Data Centres', *Annals*, vol. XXXVI (1964), pp. 592–4.

81. This assessment is based on the list of publications in the Rockets and Satellites section of *Annals*, vol. XXXVI (1964).

82. For examples of contemporary authors and later historians taking this

line see: A. Marshack, *The World in Space* (1958 – ch. 7, n. 41),
p. 160; J. M. Gavin, *War and Peace in the Space Age* (1959 – ch. 1,
n. 34), p. 222; J. T. Wilson, *IGY: the Year of the New Moons* (London:
Michael Joseph, 1961), p. 80; E. N. Hayes, *Trackers of the Skies*
(Cambridge, MA: Howard Doyle, 1968), p. 51; Green and Lomask,
*Vanguard* (n. 10), p. 185.

83.  Sullivan, *Assault* (n. 50), pp. 58–62; c.f. Sullivan, 'Year' (n. 14).
84.  H. Brooks, 'The Scientific Adviser' in R. M. Gilpin and C. Wright
     (eds), *Scientists and National Policy-Making* (New York: Columbia U.
     Press, 1964), pp. 76–7.
85.  D. L. Harvey and L. C. Ciccoritti, *U.S.-Soviet Cooperation in Space*
     (Miami: U. Miami Center for Advanced International Studies, 1974),
     p. 2.

## Chapter 9: Advice for Eisenhower

1.  J. Alsop and S. Alsop, *We Accuse! The Story of the Miscarriage of
    Justice in the Case of J. Robert Oppenheimer* (London: Gollancz,
    1955), p. 44; J. M. Gavin, *War and Peace in the Space Age* (1959
    – ch. 1, n. 34), pp. 132–5.
2.  S. P. Huntington, *The Common Defense* (New York: Columbia U.
    Press, 1961), p. 78.
3.  Ibid., p. 327.
4.  D. Caute, *The Great Fear* (1978 – ch. 1, n. 41), pp. 479–84. For
    contemporary evidence of the estrangement between government and
    science caused by McCarthyism, see the remarks by Killian and Bush
    at US Congress, House, Government Operations, Hearings: *Military
    R&D Programs* (1954 – ch. 5, n. 35), pp. 444, 454.
5.  J. R. Killian, *Sputnik, Scientists and Eisenhower* (1977 – ch. 3, n. 1),
    p. 68.
6.  D. Z. Beckler, 'The Precarious Life of Science in the White House',
    *Daedalus*, vol. 103, no. 3, 1974, p. 117.
7.  Killian, *Sputnik* (n. 5), p. 68.
8.  Ibid., pp. 88–91. In December 1956 Rabi asked Flemming to arrange
    an ODM-SAC meeting with the President by the following March. A
    meeting duly followed on 29 March 1957: DDEL – WHO: OSAST,
    Box 7, Folder: S.A.C. (3), Ltr, Rabi to Flemming, 19 December 1956;
    also DDEL – Ann Whitman File: DDE Diaries, Box 13, Folder: Office
    Memos March 1957, 'Memorandum of Conference with the President
    on March 29', 1 April 1957. A typical informal contact was the brief
    meeting between Eisenhower, Killian and Edwin Land before a session
    of the National Security Council on 17 May 1956 at which 'several'
    intelligence programmes, doubtless including the U-2, were discussed:
    DDEL – WHO: Secretary, White House – Subjects, Box 4, Folder: Dr
    Killian (1) May '56 – Nov. '57, Brig. Gen. A. J. Goodpaster, Memo.

for Record, 17 May 1956. For 'my scientists', see Killian, *Sputnik* (n. 5), p. 241.

9. John Rigden has placed on record Rabi's 'enormous respect' for Eisenhower. Rabi also told Rigden that Eisenhower's 'besetting sin, to my mind, was his modesty': J. S. Rigden, *Rabi: Scientist and Citizen* (New York: Basic Books, 1987), p. 238.

10. National Security Council, TCP Report, 'Meeting the Threat of Surprise Attack' (1955 – ch. 7, n. 29); Killian, *Sputnik* (n. 5), p. 70; M. E. Davies and W. R. Harris, *RAND's Role in the Evolution of Balloon and Satellite Observation Systems* (1988 – ch. 4, n. 32); R. M. Bissell, COHC Interview with E. Edwin, 5 June 1967, p. 38.

11. J. R. Killian: interview.

12. USNC-IGY – National Committee, Folder: Office of Defense Mobilization, Memo. D. Z. Beckler for A. S. Flemming, 'International Geophysical Year', 13 April 1954, pp. 1, 3.

13. RAND Corporation, *Project Feed Back Summary Report* R-262, 1 March 1954, (Santa Monica: RAND, 1954), p. vii.

14. R. L. Perry, *Origins of the USAF Space Program, 1945–1956* Historical Publication Series 62-24-10 (Andrews Air Force Base, MD: USAF Systems Command, 1962), pp. 36–43.

15. E. Bergaust and W. Beller, *Satellite!* (London: Scientific Book Club, 1957), pp. 34–9; NHO – W. von Braun (Biog. File), 'A Minimum Satellite Vehicle Based on Components available from missile developments of the Army Ordnance Corps', Guided Missile Development Division, Ordnance Missile Laboratories, Redstone Arsenal (mimeo), 15 September 1954. The closest contemporary *public* proposal to the Orbiter scheme assumed that a balloon satellite weighing 10 lb and inflatable to 15 feet in diameter could be placed in orbit at 200 miles altitude by 'an improved version of the V-2': I. M. Levitt, 'Geodetic Significance of a Minimum Satellite Vehicle' in *Bericht über den V. Internationalen Astronautischen Kongress*, Proc. Fifth IAF Congress, Innsbruck, August 1954 (Vienna: Friedrich Hecht, 1955.)

16. F. C. Durant: interview.

17. Bergaust and Beller, *Satellite!* (n. 15), p. 36.

18. At a briefing on 28 February 1955 von Braun told the Army Ordnance Advisory Committee that an Orbiter satellite would be 20 inches in diameter, weigh just 5 lb, have the brightness of a sixth magnitude star, and cost only $5.2 million *without* the 'logistic cost': DDEL – Furnas Papers, Box 1, Folder: Army Ordnance Advisory Committee, 1955, manuscript notes.

19. C. Mc. Green and M. Lomask, *Vanguard: a History* (1971 – ch. 6, n. 42), p. 30.

20. *PPPUS-1954*: Item 153, pp. 597–8; USNC-IGY – National Committee, Folder: DOD (Navy), text of White House Press Release on US Antarctic Expedition, 28 March 1955; *PPPUS-1957*: Item 122, pp. 511–13.

21.   DDEL – WHO: NSC Staff, OCB Central Files, Box 11, OCB 000.91, Folder: (2), Ltr, Waterman to Staats, 4 June 1955, with copies of correspondence between Kaplan, Waterman, Quarles, and Deputy Under Secretary of State Robert Murphy; Green and Lomask, *Vanguard* (n. 19), p. 30.

22.   D. R. Smith, 'They're Following Our Script: Walt Disney's Trip to Tomorrowland', *Future*, vol. 1, no. 2, 1978, p. 59.

23.   W. Kimball: interview.

24.   H. L. Goodwin, *The Science Book of Space Travel* (1954 – ch. 2, n. 14), p. 3.

25.   DDEL – Ltr, Waterman to Quarles, 13 May 1955 (n. 21).

26.   *New York Times*, 17 December 1954.

27.   See note 14.

28.   DDEL – Ann Whitman File: NSC Papers, Box 7, 'Discussion at the 283rd Meeting of the National Security Council, Thursday, May 3, 1956', 4 May 1956, p. 4. Eisenhower intended there to be no discursive minutes of NSC proceedings, as distinct from the bare record of decisions in the 'NSC Actions' series, because he did not want it to degenerate into a talking-shop. However Everett Gleason made copious notes in full view of the President during NSC meetings, which he then wrote up, usually within twenty-four hours, into extensive accounts of the discussions.

29.   DDEL – 'Discussion . . . May 3, 1956' (n. 28), p. 6 – emphasis added.

30.   NHO – Grosse Biog. File, Ltr, Grosse to Emme, 12 January 1973.

31.   L. V. Berkner, 'Science and National Strength', *Bull. Atomic Scientists*, vol. 9, no. 5, June 1953; L. V. Berkner, 'Science and Military Power', *Bull. Atomic Scientists*, vol. 9, no. 12, December 1953.

32.   L. V. Berkner, 'Earth Satellites and Foreign Policy', *Foreign Affairs*, vol. 36, no. 2, 1958, pp. 231, 221.

33.   L. V. Berkner, 'International Scientific Action' (1954 – ch. 7, n. 15), p. 575.

34.   D. K. Price, *Government and Science* (1954 – ch. 3, n. 1), p. 96.

35.   For the narrow institutional base of the science advice community, see: D. S. Greenberg, *The Politics of Pure Science* (New York: New American Library, 1971), pp. 15–16; Killian's estimate is cited without a direct reference in D. J. Kevles, *The Physicists* (1979 – ch. 3, n. 1), p. 394.

36.   D. K. Price, *The Scientific Estate* (Cambridge, MA: Harvard U. Press, 1965), p. 12 – emphasis in original.

37.   W. R. Schilling, 'Scientists, Foreign Policy, and Politics' in R. M. Gilpin and C. Wright (eds), *Scientists and National Policy-Making* (1964 – ch. 8, n. 84), p. 163.

38.   US Congress, House, Astronautics, Hearings: *Astronautics and Space Exploration* (1958 – ch. 6, n. 24), p. 1037.

39.   At a famous meeting of the National Academy of Sciences on 1 May

1958 Van Allen presented the initial measurements of the radiation belts obtained by Explorer 1 and Explorer 3. In doing so he compared them with theoretical calculations which he had made as far back as *1949*: J. A. Van Allen, 'Satellites 1958 Alpha and Gamma, High Intensity Radiation Research and Instrumentation', *IGY Satellite Report* no. 13 (Washington: National Academy of Sciences and National Science Foundation, 1961), p. 5.

40. NHO – American Rocket Society Files, 'Proc. 1st Meeting of Ad Hoc Committee on Space Flight', 17 May 1952; passage quoted from the Committee's 1953 report in: Bergaust and Beller, *Satellite!* (n. 15), pp. 32–3; American Rocket Society, Space Flight Committee, 'On the Utility of an Artificial Unmanned Earth Satellite', proposal to NSF, 24 November 1954, *Jet Propulsion*, February 1955.

41. K. R. Stehling, *Project Vanguard* (Garden City, NY: Doubleday, 1961), p. 104. That the 'science only' position of its 1954 report was adopted by the ARS only as a temporary expedient is strongly suggested by a comparison with the Society's later report, prepared in the summer of 1957 and sent to Eisenhower just after Sputnik 1: DDEL – WHO: OSAST, Alphabetical Not-Top-Secret series, Box 15, Folder: Space (Oct. 1957) (1), Report by the Space Flight Technical Committee of the ARS, 'Space Flight Program', 23 August 1957, revised 10 October 1957.

42. DDEL – Lu, Kaplan to Waterman, 14 March 1955, (n. 21).

43. J. R. Killian, 'Shaping a Public Policy for the Space Age' in L. P. Bloomfield (ed.), *Outer Space: Prospects for Man and Society* (Englewood Cliffs, NJ: Prentice-Hall, 1962), p. 184.

44. M. W. Rosen and F. L. Whipple: interviews.

45. Project RAND, J. E. Lipp, 'The Time Factor in the Satellite Program' (1946 – ch. 6, n. 39), p. 9.

46. G. W. Hoover: interview.

47. RAND Corporation, Social Sciences Division, *Proc. Conf. on Psychological Effects* (1949 – ch. 4, n. 32), p. 105.

48. RAND Corporation, Social Sciences Division, *Summary Report of Conf. on Psychological Effects* (1949 – ch. 6, n. 32), p. 15.

49. RAND Corporation, P. Kecskemeti, *The Satellite Rocket Vehicle* (1950 – ch. 4, n. 32), pp. 9, 10, 22.

50. J. C. Cooper, 'High Altitude Flight and National Sovereignty', address to Escuela Libre de Derecho, Mexico City, 5 January 1951, in US Congress, Senate, Space Sciences, Symposium: *Space Law* (85th Congress, 2nd Session, 1958), pp. 1–7.

51. O. Schachter, 'Who Owns the Universe?' in C. Ryan (ed.), *Across the Space Frontier* (1952 – ch. 4, n. 48).

52. US National Security Council, 'U.S. Scientific Satellite Program' (1955 – ch. 7, n. 29), p. 1; J. R. Killian: interview.

53. DDEL – Hagerty Papers, Box 9, Folder: Satellite Announcement, 'Questions and Answers from Meeting of Interested Representatives

from NSF, IGY, WH, OCB, NAS, State, CIA, USIA OASD-DOD', 28 July 1955.

54. A. T. Waterman, A. H. Shapley, and A. F. Spilhaus, 'ABC's of Travel in Space – Satellite Now, Moon Next?' *U.S. News & World Report*, 12 August 1955.

55. M. Caidin, *Vanguard! the Story of the First Man-Made Satellite* (New York: E. P. Dutton, 1957), pp. 265, 267; Bergaust and Beller, *Satellite!* (n. 15), p. 42.

56. J. A. Van Allen: interview.

57. US National Security Council, 'Scientific Satellite Program' (n. 52), p. 6.

58. Green and Lomask, *Vanguard* (n. 19), p. 49.

59. DDEL – Ann Whitman File: NSC Papers, Box 8, 'Discussion at the 310th Meeting of the National Security Council, Thursday, January 24, 1957', 24 January 1957, p. 6.

60. Green and Lomask, *Vanguard* (n. 19), p. 104; Stehling, *Project Vanguard* (n. 41), pp. 103–5. In Rosen's opinion, 'We could have put up the 21 lb satellite at the very beginning. . . . We were just so scared we weren't going to make it. That was a wrong decision.': interview.

61. DDEL – Ann Whitman File: NSC Papers, Box 8, 'Discussion at the 322nd Meeting of the National Security Council, Friday, May 10, 1957', 11 May 1957, pp. 5, 6.

62. US Congress, House, Armed Services, Hearings: *Investigation of National Defense Missiles* (1958 – ch. 3, n. 57), p. 4625.

63. Stehling, *Project Vanguard* (n. 41), p. 104.

64. DDEL – 'Discussion . . . May 10, 1957' (n. 61), pp. 2, 5.

65. Green and Lomask, *Vanguard* (n. 19), p. 104.

66. G. Vaeth, *200 Miles Up* 2nd edn (New York: Ronald Press, 1955), p. 237.

67. USNC-IGY – National Committee Minutes, 'Ad Hoc Meeting on Earth Satellite Program Budget', 10 November 1955; TPESP Minutes, 2nd Meeting, 21 November 1955.

68. Green and Lomask, *Vanguard* (n. 19), pp. 63–4, 104–6. McDougall cites only the Department of Defense components in the second and third of these estimates: W. A. McDougall, . . . *the Heavens and the Earth* (1985 – ch. 1, n. 47), p. 130. The Gleason Minutes describe the cost of Vanguard as having reached only $83 million by January 1957, which is inconsistent with the higher figure presented to Congress seven months earlier: DDEL – 'Discussion . . . January 24, 1957' (n. 59), p. 4.

69. US Congress, House, Appropriations: Report by Surveys and Investigations Staff, 'Project Vanguard, a Scientific Earth Satellite Program for the International Geophysical Year' in Subcommittee on Department of Defense Appropriations, Hearings: *Department of Defense Appropriations for 1960* pt 6 (86th Congress, 1st Session, 1959), p. 78. In

view of this remark it is unfortunate that the report did not establish a clear record of the successive estimates.

70. US Congress, House, Appropriations, Subcommittee on Independent Offices, Hearings: *National Science Foundation: Report on the International Geophysical Year* (85th Congress, 1st Session, 1957), p. 74.
71. DDEL – 'Discussion . . . May 10, 1957' (n. 61), p. 4.
72. M. W. Rosen: interview.
73. Ibid.
74. F. L. Whipple: interview.
75. Lipp, 'Time Factor' (n. 45), p. 1. Robert Truax repeated the $100 million estimate in a letter to the directors of the American Rocket Society in November 1951, only a few years before the US IGY Committee produced its $10 million figure: S. Thomas, *Men of Space* vol. 2 (1961 – ch. 6 n. 42), pp. 190–1.
76. US Congress, House, Armed Services, *Investigation* (n. 62), p. 4626.
77. Lipp, 'Time Factor' (n. 45), p. 4 – emphasis added.
78. Waterman, Shapley, and Spilhaus, 'ABC's of Travel in Space' (n. 54).
79. DDEL – 'Discussion . . . May 3, 1956' (n. 28), p. 3.
80. DDEL – 'Discussion . . . January 24, 1957' (n. 59), p. 5. This phrase gives Robert Cutler's summary of remarks made earlier in the meeting by Waterman, the direct record of which has been deleted.
81. DDEL – 'Discussion . . . May 3, 1956' (n. 28), pp. 3–7.
82. DDEL – Ann Whitman File: NSC Papers, Box 9, 'Discussion at the 339th Meeting of the National Security Council, Friday, October 10, 1957', 11 October 1957, p. 6.
83. DDEL – WHO: Secretary, Subjects – DOD, Box 6, Folder: Missiles & Satellites vol. I (3), Brig. Gen. A. J. Goodpaster, 'Memorandum of Conference with the President, October 8, 1957, 5 P.M.', 9 October 1957.
84. Beckler, 'Precarious Life' (n. 6); DDEL – WHO: Secretary, Subjects – Alphabetical, Box 23, Folder: S.A.C. (2), Brig. Gen. A. J. Goodpaster, 'Memorandum of Conference with the President, October 15, 1957, 11 a.m.', 16 October 1957.
85. Berkner, 'Earth Satellites' (n. 32), p. 231.
86. D. Pearson and J. Anderson, *U.S.A. – Second Class Power?* (1958 – ch. 1, n. 40), pp. 138–41.
87. US Congress, House, Armed Services, Subcommittee for Special Investigations, Hearings: *Employment of Retired Military and Civilian Personnel by Defense Industries* (86th Congress, 1st Session, 1959), pp. 862–71, 991–1006. The reliability of the General Electric engine and the absence of any comparable alternative were later confirmed by Kurt Stehling, head of Vanguard propulsion systems: *Project Vanguard* (n. 41), pp. 60–1.
88. US Congress, House, Armed Services, *Employment* (n. 87), p. 1006; D. W. Cox, *America's New Policy Makers: the Scientists' Rise to Power*

(Philadelphia: Chilton Books, 1964), p. 71.
89. L. Mallan, *Space Satellites* (Greenwich, CT: Fawcett Publications, 1958), p. 124.
90. Ibid., p. 127.
91. A. Haley, *Rocketry and Space Exploration* (Princeton: Van Nostrand, 1958), p. 155; see also J. Grey and V. Grey, *Space Flight Report to the Nation* (New York: Basic Books, 1962).
92. W. S. Bainbridge, *The Spaceflight Revolution: a Sociological Study* (New York: John Wiley, 1976), pp. 132–7.
93. Ibid., p. 137.
94. J. R. Killian: interview.

## Chapter 10: Intelligence for Eisenhower

1. E. Beard, *Developing the ICBM* (1976 – ch. 3, n. 44), pp. 153–61.
2. DeHaven – Memo. T. Gardner for Assistant Secretary of Defense (R&D), 16 February 1954, quoted in: Beard, *ICBM* (n. 1), p. 164.
3. DD-1980 – 227B: Special National Intelligence Estimate, 'Soviet Capabilities for Attack on the US Through 1957', SNIE-11-2-54, 16 February 1954, par. 30.
4. See for example: D. A. Anderton, 'Red Rockets', *Aviation Week*, 14 January 1952, and 'Red Rocket Research, Peenemuende 1951', *Aviation Age*, vol. 17, no. 2, February 1952.
5. L. Mallan, *Space Satellites* (1958 – ch. 9, n. 89), p. 126.
6. G. P. Sutton, 'Rockets Behind the Iron Curtain', *J. Am. Rocket Soc.*, vol. 23, May/June 1953; G. P. Sutton, 'Evaluation of Russian Rocket Developments', *J. Brit. Interplanetary Soc.*, vol. 13, no. 5, 1954.
7. Ibid., pp. 265–6.
8. US National Security Council, TCP Panel Report, 'Meeting the Threat of Surprise Attack' (1955 – ch. 7, n. 29).
9. Ibid.
10. Project RAND, J. E. Lipp, 'The Time Factor in the Satellite Program' (1946 – ch. 6, n. 39), p. 1.
11. DD-1986 – 000450: Memo. Eisenhower for Secretary of Defense, endorsing NSC Action No. 1433a, 16 September 1955.
12. DDEL – Ann Whitman File: NSC Papers, Box 1, Folder: Rec. of Actions by NSC 1955 (3), 'Record of Actions by the National Security Council at its Two Hundred and Sixty-Eighth Meeting, held on December 1, 1955', Action No.1484c.
13. A. J. Zaehringer, 'Soviet missile progress', *Aeronautics Digest*, vol. 71, no. 6, December 1955.
14. D. A. Anderton, 'Soviet advances force U.S. missile pace', *Aviation Week*, 12 March 1956.
15. J. T. Richelson and D. Ball, *The Ties That Bind* (1985 – ch. 5, n. 8), p. 188; J. Bamford, *The Puzzle Palace: a Report on America's Most*

*Secret Agency* (London: Penguin, 1983), p. 209. In secret testimony to the Preparedness Subcommittee hearings in November 1957 Herbert Scoville gave the numbers of Soviet missile tests at Kapustin Yar which had occurred 'since mid-1953'. This suggests, but does not prove, that American electronic monitoring of activities at Kapustin Yar began two years earlier than the generally accepted date, which has been retained in the text: DDEL – Harlow Papers, Box 1, Folder: Missile Hearings etc. 1957–58 (1), A. W. Dulles and H. Scoville, 'Comments Before the Senate Preparedness Investigating Subcommittee, November 1957', top copy of typed transcript.

16. DD-1980 – 251C: Statement of General N. F. Twining, USAF Chief of Staff, before the Subcommittee on Defense Department Appropriations of the House Committee on Appropriations, 9 February 1956.

17. Dulles and Scoville, 'Comments' (n. 15); DD-1980 – 58C: Joint Intelligence Committee, 'Semi-Annual Review of Trends in Communist Bloc Policy Including Communist China', JIC 183/3(56), 16 May 1956. Von Braun and Ordway later cited unspecified Russian sources to the effect that the first Soviet IRBM test was held in April 1956, but this was probably only the 750-mile Pobeda MRBM: W. von Braun and F. I. Ordway, *History of Rocketry & Space Travel* (1967 – ch. 1, n. 45), pp. 141–2.

18. DDEL – Quarles Papers, Box 13, Addresses as Secretary of the Air Force, vol. II, 'Address to Dean's Day Homecoming Conference, New York University, February 25, 1956.'

19. DDEL – Quarles Papers, Box 13, Addresses as Secretary of the Air Force, vol. III, 'Address before the Aviation Writers Association', 21 March 1956.

20. DD-1980 – 58C: Joint Intelligence Committee, 'Semi-Annual Review' (n. 17).

21. DDEL – WHO: Secretary, White House – Subjects, Box 4, Folder: Dr Killian (1) May 56 – Nov. 57, Brig. Gen. A. J. Goodpaster, Memo. for Record, 16 May 1956.

22. DDEL – Quarles Papers, Box 14, Addresses as Secretary of the Air Force, vol. V, 'Press Conference', 2 December 1956.

23. DDEL – Dulles and Scoville, 'Comments' (n. 15).

24. *Pravda*, 28 November 1953.

25. Sutton, 'Evaluation' (n. 6), p. 267.

26. NAS – International Relations 1957, Folder: USSR – Soviet Documents Held by US-IGY Scientists 1957, Memo. by H. Odishaw, 18 March 1957.

27. USNC-IGY – National Committee, Folder: Earth Satellite Correspondence Jan. thru Sep. 1955, Memo. P. H. Wyckoff for H. Odishaw, 'Report of Activity at CSAGI Meeting in Brussels', 26 September 1955.

28. USNC-IGY – National Committee, Folder: Satellite Correspondence January 1956, 'A Summary of Some Recent Public Information on

USSR Rocket and Satellite Developments', January 1956. For Krieger's compilations, see note 43 below.

29. USNC-IGY – National Committee, Folder: Satellite Correspondence Feb. '56, Ltr, H. Odishaw to L. V. Berkner, 8 February 1956.

30. NAS – International Relations 1957, Folder: IGY General, Ltr, R. Fraser to L. V. Berkner, 23 July 1957.

31. DD-1980 – 58C: Joint Intelligence Committee, 'Semi-Annual Review' (n. 17), p. 6.

32. S. Alsop, *The Center* (London: Hodder and Stoughton, 1968), pp. 243–4.

33. For example: H. E. Ross, 'Wide Peace Use Seen for Earth Satellite', *Aviation Week*, 27 April 1953; W. von Braun, 'The Baby Space Station' (1953 – ch. 4, n. 55).

34. US National Security Council, 'U.S. Scientific Satellite Program', (1955 – ch. 7, n. 29), pp. 1–2, 3.

35. DDEL – Hagerty Papers, Box 9, Folder: Satellite Announcement, A. Waterman, 'Notes on a Meeting with the President on July 27, 1955, Preparatory to a Meeting with Hagerty on July 28, 1955', n.d.

36. Memo. Maj. Gen. L. Simon for H. J. Stewart, 15 August 1955, in US Congress, House, Appropriations, Report: 'Project Vanguard' (1959 – ch. 9, n. 69), pp. 58–9.

37. DDEL – Ann Whitman File: NSC Papers, Box 7, 'Discussion at the 283rd Meeting of the NSC, May 3, 1956', 4 May 1956, p. 5.

38. DDEL – Furnas Papers, Box 2, Folder: Def. Dept. of – General Corr. 1955, Ltr, P. A. Smith to H. J. Stewart, 28 September 1955.

39. DDEL – Ann Whitman File: NSC Papers, Box 8, 'Discussion at the 310th Meeting of the NSC, January 24, 1957', 24 January 1957; DDEL – ibid., 'Discussion at the 322nd Meeting of the NSC, May 10, 1957', 11 May 1957.

40. DDEL – Ann Whitman File: Administration, Box 37, Folder: US Satellites, A. W. Dulles, untitled second draft of possible government statement in response to Sputnik 1, n.d. but c. 7 October 1957, pp. 1–2 – emphasis added.

41. D. D. Eisenhower, *Waging Peace* (London: Heinemann, 1966), p. 206.

42. R. S. Cline, *Secrets, Spies and Scholars* (Washington: Acropolis, 1972), pp. 124–5. The NSC meeting recollected by Cline may have been that held on 24 September 1957, which was attended by an unnamed CIA official together with Dulles: DDEL – Ann Whitman File: NSC Papers, Box 9, Folder: Records of Action by NSC, 1957 (3), 'Record of Actions by the National Security Council at its Three Hundred and Thirty-Seventh Meeting, held on September 24, 1957'. But Cline now feels that the meeting he described 'was at least two or three months earlier': personal communication.

43. Similarly, the first RAND compilation of Soviet writings on astronautics, which was circulated in June 1956, referred only to 'a definite

competence in interplanetary communications' (p. 19), and did not include the estimate that the Russians might launch a satellite in 1957 which was added in the sequel a year later: RAND Corporation, F. J. Krieger, *A Casebook on Soviet Astronautics RM-1760* (Santa Monica: RAND, 1956); RAND Corporation, F. J. Krieger, *A Casebook on Soviet Astronautics* 2nd edn RM-1922 (Santa Monica: RAND, 1957).

44.  DD-1978 – 146A: Advisory Group on Special Capabilities, 'Memo. for the Assistant Secretary of Defense (R&D), Subject: VANG and REDSTONE', 22 June 1956; ibid., Memo. E. V. Murphree for Deputy Secretary of Defense, 'Use of the JUPITER Re-entry Test Vehicle as a Satellite', 5 July 1956.

45.  This idea was published by Albert Parry and by Kenneth Gatland in the July 1957 issues of *Missiles and Rockets* and *Spaceflight*, respectively, though Gatland seems to have forgotten to adjust his anniversary prediction to the post-revolutionary calendar.

46.  J. M. Gavin, *War and Peace in the Space Age* (1959 – ch. 1, n. 34), p. 29.

47.  B. Harvey, *Race Into Space* (1988 – ch. 1, n. 32), p. 27.

48.  D. Baker, *The Rocket: the History and Development of Rocket and Missile Technology* (London: New Cavendish Press, 1978), p. 120.

49.  It is probably some phase of this relatively technical decision that Korolev is said (without a reference to any source) to have represented as 'the Central Committee finally endors[ing] the project' only in the summer of 1957: J. Oberg, *Red Star in Orbit* (New York: Random House, 1981), p. 30.

50.  Harvey, *Race Into Space* (n. 47), p. 29.

## Chapter 11: Eisenhower's Early Space Policy

1.  M. E. Davies and W. R. Harris, *RAND's Role in the Evolution of Balloon and Satellite Observation Systems* (1988 – ch. 4, n. 32), pp. 57–9.

2.  J. T. Richelson and D. Ball, *The Ties That Bind* (1985 – ch. 5, n. 8), pp. 252–3.

3.  C. Mc. Green and M. Lomask, *Vanguard: a History* (1971 – ch. 6, n. 42), pp. 30–1.

4.  US National Security Council, 'U.S. Scientific Satellite Program', (1955 – ch. 7, n. 29), p. 6 par. 12.

5.  Ibid., p. 5 par. 10, p. 6 par. 11. The language in the first of these passages reflects the impression which had been left with the decision-makers that the project would be a version of the Redstone-based Orbiter.

6.  DDEL – Hagerty Papers, Box 9, Folder: Satellite Announcement, NSF and NAS Press Release, 'Plans for Construction of Earth Satellite Vehicle Announced', 29 July 1955, p. 1.

7.  DDEL – Furnas Papers, Box 15, Speeches, 'Around the World in 90

Minutes', address before the American Rocket Society, 25 September 1956, pp. 4–5.
8.   DDEL – Quarles Papers, Box 14, Addresses vol. V, 'Press Conference', 2 December 1956. See also DDEL – NSF/NAS Press Release (n. 6), and A. T. Waterman, A. H. Shapley and A. F. Spilhaus, 'ABC's of Travel in Space', *U.S. News & World Report*, 12 August 1955.
9.   W. A. McDougall, . . . *the Heavens and the Earth* (1985 – ch. 1, n. 47), pp. 122–3.
10.  Green and Lomask, *Vanguard* (n. 3), p. 49.
11.  US Congress, House, Appropriations, Report: 'Project Vanguard' (1959 – ch. 9, n. 69), p. 58.
12.  C. C. Furnas, 'Why Did U.S. Lose the Race?' *Life*, 21 October 1957, p. 22.
13.  M. W. Rosen: interview.
14.  Secretary Wilson's intimation to the contrary, at a 1956 meeting of the National Security Council, was mentioned in Chapter 9. But that was more an expression of ill-advised resentment at having to fund the internationally public and scientific Vanguard project through the Department of Defense, and of concern for the Jupiter IRBM project, than a statement of fact about the state of the Redstone missile project in the preceding summer.
15.  US Congress, House, Armed Services, Hearings: *Investigation of National Defense Missiles* (1958 – ch. 3, n. 57), p. 4150; for the overall sequence of events, see US Congress, House, Appropriations, 'Project Vanguard' (n. 11).
16.  DD-1978 – 146A: Advisory Group on Special Capabilities, 'Memo. for the Assistant Secretary of Defense (R&D), Subject: VANG and REDSTONE', 22 June 1956. The problem with the advice on tracking facilities was that the memorandum was based on a meeting of the Group which took place on 23 and 24 April 1956 and the first successful tests of the JPL Microlock system were conducted at Earthquake Valley, California, between 14 and 25 May. A second Microlock unit was then built at Cape Canaveral to support the Jupiter C test firings. Nevertheless it seems extraordinary that at least until July 1956 and possibly even later the administration continued to be told that the United States had no satellite-tracking capability whatsoever, at the very moment when that was ceasing to be the case.
17.  Ibid., Memo. E. V. Murphree for Deputy Secretary of Defense, 'Use of the JUPITER Re-entry Test Vehicle as a Satellite', 5 July 1956.
18.  F. I. Ordway and M. R. Sharpe, *The Rocket Team* (1979 – ch. 3, n. 38), p. 380.
19.  US Congress, House, Government Operations, Hearings: *Research and Development* pt 1 (85th Congress, 2nd Session, 1958), pp. 253–4.
20.  DDEL – NASA Papers, Box 1, Folder: Working Group on Certain Aspects of the Scientific Earth Satellite – Minutes 1956–58, 'Memorandum of Meeting etc., June 13, 1957', 13 June 1957, p. 1.
21.  DDEL – ibid., 'Memorandum of Meeting etc., June 17, 1957', 17 June

1957, attachment.
22. DDEL – ibid., 'Memorandum of Meetings etc., October 5 and 7, 1957', 8 October 1957.
23. DDEL – WHO: OSANSA, NSC Briefings, Box 7, Folder: Earth Satellites (1), D. A. Quarles, 'Memorandum for the President, Subject: Earth Satellite, 7 October, 1957', pp. 2–5 – emphases added.
24. DDEL – WHO: OSANSA, NSC Briefings, Box 7, Folder: Earth Satellites (1), unsigned minutes, 'Conference in the President's Office, 8.30 a.m., Oct. 8, 1957', p. 1.
25. DDEL – WHO: NSC Staff, Executive Secretary, Subjects, Box 14, Folder: OCB (1), Presidential Executive Order, 3 September 1953.
26. DDEL – Ann Whitman File: Administration, Box 29, Memo. C. D. Jackson for R. Cutler, 3 April 1953.
27. DD-1977 – 306A: Memo. J. Lay, NSC Executive Secretary, for NSC Staff, 'The Foreign Information Program and Psychological Warfare Planning', 15 March 1955, appendix, p. 10, passage quoted from an OCB report to the NSC of 21 July 1954.
28. DDEL – WHO: NSC Staff, Executive Secretary, Subjects, Box 14, Folder: OCB (3), Ltr, Eisenhower to Herter, 25 February 1957; DDEL – WHO: OSANSA, Special Assistant – Chronological, Box 4, Folder: May 1957 (2), Draft Ltr, Cutler to Eisenhower, resigning as OCB Vice-Chairman, 16 May 1957. At the time of Sputnik 1 the OCB comprised Cutler, Under Secretary of State Christian Herter (chairman), Deputy Secretary of Defense Donald Quarles, DCI Allen Dulles, and the director of the US Information Agency, Arthur Larson.
29. US National Security Council, 'Scientific Satellite Program' (n. 4), p. 3 par. 6 and Annex B.
30. DDEL – WHO: NSC Staff, OCB Central Files, Box 11, OCB 000.91, Folder: 1 (2), Ltr, Waterman to Staats, 4 June 1955.
31. DDEL – ibid., 'Memorandum of Meeting, Ad Hoc Working Group on Public Information Aspect of NSC 5520, Meeting No. 3, July 26, 1955', 26 July 1955; 'Memorandum of Meeting etc., July 28, 1955', 29 July 1955; ibid., Folder 1 (3): 'Memorandum of Meeting etc., August 3, 1955', 3 August 1955.
32. USNC-IGY – TPESP, Folder: Earth Satellite Corresp. – Jan. thru Sept. 1955, Ltr, Kaplan to Hugh [Odishaw], 18 July 1955.
33. NAS – International Relations 1955, Folder: IGY, U.S. National Committee – General, Ltr, Berkner to Nicolet, 22 July 1955.
34. DDEL – WHO: NSC Staff, OCB Central Files, Box 11, OCB 000.91, Folder: 1 (1), Draft Ltr, Kaplan to Chapman, 19 July 1955; DDEL – Hagerty Papers, Box 9, Folder: Satellite Announcement, NSF-NAS Press Release, 29 July 1955, attachment: Ltr, Kaplan to Chapman, 26 July 1955; DDEL – ibid., A. T. Waterman, 'Notes on a Meeting with the President on July 27, 1955, Preparatory to a Meeting with Hagerty on July 28, 1955', n.d. The displacement of Chapman in London, the nominal addressee of 'Kaplan's' letter, by Nicolet in Brussels, was

due to 'objections' raised at the Working Group to 'the possibility of a British dateline on the announcement': E. B. Staats, 'Memorandum for the Chairman, Operations Coordinating Board', 25 July 1955, attached to 'Memorandum of Meeting etc., July 28, 1955' (n. 31).

35. DDEL – WHO: NSC Staff, OCB Central Files, Box 11, OCB 000.91, Folder: 1 (2), 'Draft of a Suggested Press Release by the National Academy of Sciences and the National Science Foundation', 19 July 1955.

36. DDEL – Hagerty Papers, Box 9, Brig. Gen. A. J. Goodpaster, 'Memorandum of Conference with the President, July 27, 1955 11.45 A.M.', 27 July 1955.

37. DDEL – 'Memorandum of Meeting etc., August 3, 1955' (n. 31), Item 1.

38. USNC-IGY – Executive Committee Correspondence, Folder: August 1955, Ltr, Kaplan to Waterman, 3 August 1955.

39. C. C. Adams, *Space Flight* (1958 – ch. 1, n. 38), p. 120.

40. J. L. Gaddis, *Strategies of Containment* (Oxford: Oxford U. Press, 1982), p. 155.

41. DDEL – 'Memorandum of Meeting etc., August 3, 1955' (n. 31), Item 4; DDEL – WHO: Secretary, Subjects – Alphabetical, Box 22, Folder: OCB vol. I (6), 'Major Items of Current Concern to the O.C.B.' 19 September 1955.

42. DDEL – WHO: NSC Staff, OCB Central Files, Box 2, OCB 000.7, R. Hirsch, 'Staff Study on Coordination of Public Information on U.S. IGY Programs, 1957–58', n.d. but c. December 1955 (attached to Memo. J. W. Joyce for R. Hirsch, 20 December 1955), p. 3.

43. DDEL – WHO: NSC Staff, OCB Central Files, Box 11, OCB 000.91, Folder: 1 (3), Agenda of Meeting for Representatives of Government Agencies in the Satellite Program, 18 January 1956.

44. DDEL – 'Staff Study' (n. 42); DDEL – WHO: NSC Staff, OCB Central Files, Box 11, OCB 000.91, Folder: 1 (4), 'Terms of Reference for Working Group on Certain Aspects of NSC 5520', drafts dated 2, 3 and 20 February 1956, with comments and notes on Defense and NSF positions; ibid., Memo. R. Hirsch for E. Staats, 'Background of Interdepartmental Coordination of NSC 5520', 16 February 1956.

45. DDEL – ibid., extract from Minutes of 337th Meeting of the OCB, Item (e), 13 April 1956.

46. DDEL – NASA Papers, etc. (n. 20) 'Memorandum of Meeting etc., May 18, 1956', 21 May 1956; ibid., 'Memorandum of Meeting etc., August 28, 1956', 28 August 1956.

47. USNC-IGY – Antarctic Program, Folder: May 1954, 'USNC-IGY Antarctic Committee Meeting', 13 October 1954, p. 1. In the 1954–5 Antarctic season the US IGY Committee was duly represented on the USS *Atka* by Paul Humphrey, a scientist employed by the US Weather Bureau: W. Sullivan, *Quest for a Continent* (London: Secker & Warburg, 1957), p. 325.

48. USNC-IGY – National Committee, Folder: DOD (Navy), text of White House Press Release on US Antarctic Expedition, 28 March 1955; see also: *PPPUS-1958*, p. 866.
49. USNC-IGY – 'Antarctic Committee Meeting' (n. 47), p. 4.
50. DDEL – Memo. Joyce for Hirsch (n. 42), 20 December 1955.
51. DDEL – 'Memorandum of Meeting etc., July 28, 1955' (n. 31), p. 2.
52. US Air Force, 1st Air Division (Meteorological Survey), 'Final Report: Project 119L', pp. 20, 24, 25, 17–18, in R. W. Sørdahl, *Den store hvite hvalen: Amerikansk etteretning fra Gardermoen 1956* PRIO Report No. 5 (Oslo: International Peace Research Institute, 1984).
53. NAS – International Relations 1956, International Geophysical Year, U.S. National Committee – General, 'Memorandum on Air Force Balloons and the IGY', p. 1.
54. NAS – ibid., pp. 5, 6–7.
55. *Time*, 23 April 1956; *Washington Post*, 31 May 1956.
56. 'Race for "Moon" Spurs U.S.', *Christian Science Monitor*, 4 May 1956.
57. USNC-IGY – TPESP, Satellite Correspondence, Folder: May 1956, Inter-Office Memo. G. F. Schilling for H. Odishaw, Subject, Visual Observation Program-ESP, 2 May 1956.
58. USNC-IGY – TPESP, Satellite Correspondence, Folder: June 1956, Ltr, Spitz to Odishaw, 2 June 1956, attachment, 'Statement by Armand Spitz Regarding His Montreal Talk' – emphasis added. Spitz's explanation was apparently accepted, since he was later called in to coordinate the Moonwatch search for Sputnik 1.
59. F. L. Whipple: interview.
60. USNC-IGY – TPESP: Satellite Correspondence, Folder: July 1956, Note to Files, Subject: Meeting of Dr. Kaplan and Dr. Bronk at NAS, July 27, 1956, 2.00 p.m, p. 1; DDEL – WHO: NSC Staff, OCB Central Files, Box 11, OCB 000.91, Folder: 1 (7), Memo. 'The IGY Satellite Program: Policies and Responsibilities', attached to Ltr, Kaplan to Holaday, 13 July 1956, all enclosed with Ltr, Odishaw to Furnas, 31 August 1956, p. 3.
61. DDEL – Memo. 'The IGY Satellite Program' (n. 60), pp. 2, 3.
62. USNC-IGY – TPESP: Folder: Sat. Corr. July 56, Memo. to Files from J. Kaplan, Subject: Telephone Call from Waterman to Kaplan, 25 July 1956 (12.15 p.m).
63. USNC-IGY – TPESP: Minutes, Seventh Meeting, 5 September 1956, attachment, Ltr, Porter to Kaplan, 17 July 1956.
64. NAS – International Relations 1956, Folder: International Geophysical Year, U.S. National Committee Panels – Rocketry, Earth Satellite, Ltr, Kaplan to Waterman, 24 July 1956, p. 1.
65. DDEL – WHO: NSC Staff, OCB Central Files, Box 11, OCB 000.91, Folder: 1 (7), Ltr, Odishaw to Furnas, 31 August 1956.
66. DDEL – WHO: OSANSA, NSC Briefings, Box 7, Folder: Earth Satellites (1), Ltr, Rabi to Flemming, 10 October 1956, p. 3.

67. NAS – International Relations 1956, Folder: International Geophysical Year, U.S. National Committee Panels – Rocketry, Earth Satellite, R. C. Peavey, Memo. to Files, 12 December 1956.

68. DDEL – WHO: NSC Staff, OCB Central Files, Box 11, OCB 000.91, Folder: 1 (9), Memo. C. E. Johnson for E. Staats, 20 December 1956.

69. DDEL – WHO: OSANSA, OCB – Subjects, Box 4, Folder: Missile Publicity (2), Killian, Memo. for the President, Subject: Publicity Concerning U.S. Earth Satellite Tests, 6 December 1957. Killian's proposal for this further policy review had been accepted at the previous day's meeting of the NSC, at which the President had once again raised the idea of witholding publicity about attempted satellite launchings until after they were successful: DD-1987 – 1666: Record of Actions by the NSC, 347th Meeting, 5 December 1957, NSC Action No. 1822. On 6 December, after Eisenhower's comment, the first attempted launching of a small Vanguard test satellite ended disastrously on the launch-pad before the eyes of the world press.

70. DDEL – Ann Whitman File: NSC Papers, Box 8, 'Discussion at the 310th Meeting of the National Security Council, Thursday, January 24, 1957', 24 January 1957, p. 3 – emphases added.

71. DDEL – Ann Whitman File: NSC Papers, Box 8, 'Discussion at the 322nd Meeting of the National Security Council, Friday, May 10, 1957', 11 May 1957, p. 5.

72. US Congress, Senate, Appropriations, Hearings: *The Supplementary Appropriations Bill, 1958* (85th Congress, 1st Session, 1957), p. 98.

73. DDEL – NASA Papers, etc. (n. 20) 'Memorandum of Meeting etc., October 5 and 7, 1957', 8 October 1957, p. 2.

74. DDEL – ibid., 'Memorandum of Meeting etc., October 15, 1957', 15 October 1957, Item 1.b; ibid., 'Memorandum of Joint Meetings of OCB Working Group on Nuclear Energy and OCB Working Group on Certain Aspects of NSC 5520, October 21, 1957', 22 October 1957, attachment: draft statements prepared by US IGY Committee, 18 October 1957.

75. DDEL – ibid., 'Memorandum of Joint Meetings etc., October 21, 1957', 22 October 1957, p. 2 – emphasis added.

76. DDEL – ibid., 'Memorandum of Meeting etc., February 11, 1958', 12 February 1958, pp. 2–3.

77. DDEL – ibid., 'Memorandum of Meeting etc., October 28, 1957', 30 October 1957, p. 2.

78. DDEL – ibid., 'Memorandum of Meeting etc., November 5, 1957', 6 November 1957, p. 2; ibid., 'Memorandum of Meeting etc., January 24, 1958', 29 January 1958, p. 2.

79. DDEL – ibid., 'Memorandum of Meeting etc., November 21, 1957', 25 November 1957, p. 3.

80. DDEL – ibid., 'Memorandum of Meeting etc., December 10, 1957', 11 December 1957, pp. 2–3.

81. DDEL – ibid., p. 3.

82. I. Harris and R. Jastrow, 'Demise of Satellites 1957α1 and 1957β', *Annals*, vol. XII pt I, 1960; see also L. G. Jacchia, 'The Descent of Satellite 1957β1', ibid.

83. DD-1985 – 002152: OCB Weekly Activity Report, 20 January 1958, Item 15.

84. DDEL – Ann Whitman Diary File, Box 9, Folder: Jan. 1958 (1), 'Memorandum on Telephone Calls Between Brigadier General Andrew J. Goodpaster and James C. Hagerty in Augusta, Georgia, Friday Afternoon and Evening, January 31, 1958 and Saturday Morning, February 1, 1958', pp. 12–13.

85. DDEL – NASA Papers, etc. (n. 20), 'Amendment to Memorandum of Meeting, January 29, 1958' circulated by Richard Hirsch, 24 February 1958, attachment: Ltr, J. Truesdale (US IGY Committee Office of Information) to R. Hirsch, 20 February 1958.

86. DDEL – ibid., 'Memorandum of Meeting etc., December 10, 1957', 11 December 1957, p. 2.

87. DDEL – ibid., 'Memorandum of Meeting etc., January 24, 1958', 29 January 1958, p. 1.

88. DDEL – ibid., 'Memorandum of Meeting etc., February 11, 1958', 12 February 1958, pp. 1–2.

89. DDEL – ibid., 'Memorandum of Meeting etc., March 7, 1958', 10 March 1958, *et sqq.*

90. DDEL – Memo. 'The IGY Satellite Program' (n. 60), p. 3.

91. DDEL – Ann Whitman File: NSC Papers, Box 8, 'Discussion . . . May 10, 1957' (n. 71); ibid., Box 9, 'Discussion at the 339th Meeting of the National Security Council, Thursday, October 10, 1957', 11 October 1957.

92. DDEL – WHO: OSANSA, OCB Admin., Box 1, Folder: Chron. F. M. Dearborn Nov-Dec 1957 (4), F. M. Dearborn, 'Memorandum to the Chairman, Operations Coordinating Board', 18 December 1957.

93. DDEL – ibid., Box 5, Folder: Special OCB Cttee (6) Chron. File, 'Projects to Regain the Initiative', 24 January 1958; the last item was added in pencil on Dearborn's own copy: ibid., F. M. Dearborn Series, Folder: Jan. 58 (1).

94. DDEL – Dearborn, 'Memorandum to the Chairman' (n. 92); DDEL – 'Projects' (n. 93).

95. R. C. Hall, *Lunar Impact: a History of Project Ranger* (Washington: NASA, 1977), pp. 4–6.

96. DDEL – Dearborn, 'Memorandum to the Chairman' (n. 92).

97. DDEL – WHO: OSANSA, OCB Admin., Box 5, Folder: Spec. OCB Cttee, F. M. Dearborn (2), F. M. Dearborn, 'Memorandum to the Special OCB Committee', 14 February 1958; Hall, *Impact* (n. 95), pp. 6–10. The tremendous pressure of public expectation on the engineers involved in the Pioneer launches was vividly portrayed by two of their number: T. J. Gordon and J. Scheer, *First Into Outer Space* (New York: St Martin's Press, 1959).

98. Hall, *Impact* (n. 95), p. 10; see also U.S.S.R. Academy of Sciences (trans. J. B. Sykes), *The Other Side of the Moon* (Oxford: Pergamon, 1960).
99. BCIGY – Minutes of the Artificial Satellite Subcommittee, Informal Meeting to Discuss Artificial Satellites, 2 May 1958 – emphasis added.
100. Ibid.
101. DDEL – NASA Papers, etc. (n. 20), 'Memorandum of Meeting etc., April 16, 1958', 22 April 1958, *et sqq.*
102. DDEL – WHO: OSANSA, Special Assistant – Subjects, Box 6, Folder: N.S.F. (3), Memo. A. T. Waterman for E. B. Staats, plus Enclosure 7, 16 December 1957. See also D. S. Greenberg, 'Mohole: the Project that Went Awry', *Science*, vol. 143, 10, 17 and 24 January 1964.
103. NAS – Ltr, Kaplan to Waterman, 24 July 1956 (n. 64).
104. J. R. Killian, 'Shaping a Public Policy for the Space Age' (1962 – ch. 9, n. 43), p. 184 – emphasis in original.
105. H. Odishaw, 'International Cooperation in Space Science' in L. P. Bloomfield (ed.), *Outer Space* (1962 – ch. 9, n. 43), p. 111.
106. H. Odishaw, 'International Geophysical Year' pt 2, *Science*, vol. 129, 2 January 1959.
107. NAS – International Relations 1956, Folder: International Geophysical Year, U.S. National Committee Panels – Rocketry, Earth Satellite, Ltr, Waterman to Kaplan, 6 September 1956 – emphases added.
108. V. Van Dyke, *Pride and Power* (1965 – ch. 3, n. 45), p. 14.
109. C. C. Alexander, 'Part One – Research' (1966 – ch. 6, n. 55), p. 82.
110. J. M. Logsdon, *The Decision to Go to the Moon: Project Apollo and the National Interest* (Cambridge, MA: MIT Press, 1970), p. 13; see also McDougall, *Heavens* (n. 9), pp. 122–3. McDougall's argument is further discussed in Chapter 13.
111. McDougall mentions the Working Group, but only in its third phase, which began when NASA was created in the autumn of 1958. His comments give the impression that it was first set up at this point, three years after it actually was: *Heavens* (n. 9), pp. 183–4.
112. *Aviation Week*, 14 October 1957.
113. US Congress, Senate, Armed Services, Hearings: *Inquiry into Satellite and Missile Programs* (1958 – ch. 1, n. 11), p. 1635.
114. P. J. Klass, *Secret Sentries in Space* (New York: Random House, 1971); G. M. Steinberg, *Satellite Reconnaissance: the Role of Informal Bargaining* (New York: Praeger, 1983); P. B. Stares, *Space Weapons and U.S. Strategy: Origins and Development* (Beckenham: Croom Helm, 1985).
115. McDougall, *Heavens* (n. 9), pp. 120–1.
116. J. Holmes, *America on the Moon* (1962 – ch. 7, n. 31), p. 49.
117. US Congress, House, Appropriations, 'Project Vanguard' (n. 11).
118. Green and Lomask, *Vanguard* (n. 3), p. 36 – emphasis added; see also K. R. Stehling, *Project Vanguard* (1961 – ch. 9, n. 41), pp. 50–4.

119. McDougall treats the conjecture mooted by Green and Lomask as an established fact: *Heavens* (n. 9),.p. 122.

## Chapter 12: Post-Mortem

1. US Congress, Senate, Armed Services, Hearings: *Inquiry into Satellite and Missile Programs* (1958 – ch. 1, n. 11 – hereafter *ISMP*), p. 2039.
2. *ISMP* (n. 1) 'Statement of the Preparedness Investigating Subcommittee,' 23 January 1958, p. 2427.
3. *ISMP*, 'Statement etc.' (n. 2), p. 2428.
4. L. B. Johnson, '"State of the Nation" Address' to Democratic senators on 7 January 1958, *New York Times*, 8 January 1958; for 'national policy' see ibid.
5. LBJL – Senate Papers: Siegel Papers, Box 407, Memo. Siegel for Johnson, 6 April 1956.
6. LBJL – Senate Papers: Reedy Office Files, Box 420, Memo. Reedy for Johnson, 18 March 1957 – emphases in original.
7. LBJL – Senate Papers: Preparedness Investigating Subcommittees, Box 346, Memo. A. P. Bryant for Johnson, n.d. but c. January 1953; LBJL ibid., Siegel Papers, Box 407, Memos. Siegel to Johnson, 2 and 7 February 1957. (On delayed deliveries of guided missiles, see also ibid., Box 346, Memo. Reedy for Johnson, 22 January 1952.)
8. LBJL – Memo. Bryant for Johnson (n. 7); LBJL – Senate Papers: Siegel Papers, Box 407, Folder: Memoranda Jan. 1 '57 thru [*sic*], Memo. B. Mooney for Johnson, 15 February 1956.
9. NHO – G. W. Siegel, OH Interview with E. M. Emme, 8 July 1976. In the same interview, however, Siegel also claimed that the Preparedness Subcommittee had produced one report on missiles before the sputniks. The author has been unable to trace any such report, unless Siegel was referring to: US Congress, Senate, Armed Services, Report: *Guided Missiles in Foreign Countries* (85th Congress, 1st Session, 1957), which was hastily compiled for the Subcommittee in April that year by Congressional researcher Eilene Galloway (E. M. Galloway: interview). But as the title shows, that report was not concerned with the American missile programme as such.
10. LBJL – Senate Papers: Reedy Office Files, Box 421, Folder: Memos. – Others, Memo. J. H. Rowe for Johnson, 3 July 1957.
11. DDEL – WHO: OSANSA, Special Assistant – Subjects, Box 7, Folder: S A C (3) March–Oct. 1957, handwritten notes, probably by Robert Cutler, on meetings of ODM-SAC scientists with Eisenhower and McElroy, 15 October 1957, p. 5.
12. LBJL – Senate Papers: Reedy Office Files, Box 420, Folder: Reedy: Memos. - October 1957, Ltr and Memo. Reedy for Johnson, 17 October

1957, p. 1. McDougall discusses this memorandum and Reedy's covering letter at some length, but mistakenly attributes the memorandum itself to Charles Brewton, who had stimulated Reedy to write it: W. A. McDougall, . . . *the Heavens and the Earth* (1985 – ch. 1, n. 47), pp. 141–50.

13.　LBJL – Memo. Reedy for Johnson (n. 12), pp. 3, 4 – emphasis in original.

14.　*Time*, 28 October 1957.

15.　LBJL – Memo. Reedy for Johnson (n. 12), p. 4.

16.　LBJL – Senate Papers: Reedy Office Files, Box 421, Memo. Reedy for Johnson, n.d. (beginning 'In view of the shortness of time . . . ').

17.　LBJL – Senate Papers: Preparedness Investigating Subcommittees, Box 356, Folder: Teller, Edward, Dr., Memo. Siegel for Johnson, 30 October 1957, p. 1. The article was not identified, but was probably E. Teller, 'We must win the H-war – before it starts!' New York *Herald Tribune*, 13 October 1957 (magazine pp. 8–10). Teller followed this by publishing the text of a speech made on 2 November: 'We Have Suffered a Very Serious Defeat', *U.S. News & World Report*, 15 November 1957.

18.　G. E. Reedy: interview.

19.　Ibid. Teller's selection as the opening witness may also have owed something to his having been appointed in mid-October by Secretary of the Air Force James Douglas to chair a special advisory committee of the USAF Scientific Advisory Board tasked with drawing up 'a line of positive action' for the Air Force in space: C. C. Alexander, 'Part One – Research' (1966 – ch. 6, n. 55), p. 73.

20.　*ISMP* (n. 1), p. 2092; c.f. F. Zwicky, 'A Stone's Throw Into the Universe: a Memoir' in R. C. Hall (ed.), *History of Rocketry and Astronautics* (1977 – ch. 3, n. 51).

21.　C. C. Furnas, 'Why Did U.S. Lose the Race?' *Life*, 21 October 1957; Furnas referred to an interest in satellites amongst scientists and military officers 'more than a decade ago' on the basis of which 'a U.S. satellite could have been orbiting the earth as early as 1955'.

22.　O. M. Gale, 'Post-Sputnik Washington from an Inside Office', reprinted from: *Bulletin of the Cincinnati Historical Society*, volume 43, Winter 1973, no. 4, p. 232.

23.　*ISMP* (n. 1), pp. 2, 3.

24.　Ibid., p. 88.

25.　Ibid., p. 193.

26.　Ibid., p. 584.

27.　Ibid., p. 283.

28.　Ibid., pp. 1028 ff.

29.　Ibid., p. 236.

30.　Ibid., p. 317.

31.　The only full Congressional report, written in January 1959, was US Congress, House, Appropriations, Report: 'Project Vanguard, a

Scientific Earth Satellite Program for the International Geophysical Year' (1959 – ch. 9, n. 69).
32. US Congress, House, Armed Services, Hearings: *Investigation of National Defense Missiles* (1958 – ch. 3, n. 57), p. 4613; see also J. R. Killian, *Sputnik, Scientists and Eisenhower* (1977 – ch. 3, n. 1), p. 76.
33. *ISMP*, 'Statement etc.' (n. 2), p. 2429.
34. *ISMP* (n. 1), p. 1638.
35. Johnson, '"State of the Nation" Address' (n. 4).
36. D. Cole, 'Response to the "Panama Hypothesis"', *Astronautics*, vol. 10, June 1961; c.f. W. von Braun, 'Space Superiority' (1953 – ch. 4, n. 58) and 'Station In Space' (1953 – ch. 4, n. 54).
37. Johnson, '"State of the Nation" Address' (n. 4) – emphases added.
38. *ISMP* (n. 1), p. 23 ff.
39. *ISMP*, 'Statement etc.' (n. 2), p. 2429.
40. *ISMP* (n. 1), p. 1320; *Time*, 16 December 1957.
41. *ISMP* (n. 1), p. 1976. According to Reedy (interview), it was partly because of 'a very peculiar inner life, in which he would sort of play games with himself', that Johnson 'would never come out, if he could possibly help it, with any clear and definite statement about where he was going.'
42. US Congress, Joint Committee on the Investigation of the Pearl Harbor Attack, Hearings and Reports: *Investigation of the Pearl Harbor Attack* (80th Congress, 1st Session, 1947). The remarks made here are not intended to imply that the Congressional Pearl Harbor Inquiry of 1945–6 did a satisfactory job of uncovering the historical truth about the events which it investigated, but merely that it carried out a far more thorough and comprehensive historical investigation than was achieved by the Preparedness Subcommittee hearings after the first sputniks. For an account of the Japanese attack on Pearl Harbor and an assessment of subsequent official enquiries in the United States, see: J. Toland, *Infamy: Pearl Harbor and its Aftermath* (London: Methuen, 1982).
43. One historian has stated that 'the total printed testimony ran to more than seven thousand pages' – E. C. B. Schoettle, 'The Establishment of NASA' (1966 – ch. 7, n. 46), p. 223; but that was a mistaken reading of Johnson's résumé of the proceedings, which had actually referred to 'pages' of typescript from the stenographic record – *ISMP*, 'Statement etc.' (n. 2), p. 2427.
44. LBJL – E. M. Galloway, OH Interview with M. L. Gillette, 18 May 1982, p. 1.
45. *ISMP*, 'Statement etc.' (n. 2), p. 2428.
46. US Congress, House, Armed Services, *National Defense Missiles* (n. 32), pp. 4283 ff., 4641, 4297.
47. US Congress, House, Astronautics, Hearings: *Astronautics and Space Exploration* (1958 – ch. 6, n. 24), pp. 1502–4, 54.

48. Ibid., pp. 149, 63–4, 275.
49. Ibid., pp. 351, 665, 904.
50. Ibid., pp. 279, 293.
51. NHO – E. M. Galloway, OH Interview with E. M. Emme and A. F. Roland, 9 April 1974.
52. LBJL – Senate Papers: Reedy Office Files, Box 421, Folder: Reedy Memos. Nov. 1957, 2 of 2, 'Outline of Functions for Scientific Advisers', 22 November 1957. Vannevar Bush had originally recommended Furnas for the post of senior scientific adviser for the Subcommittee hearings: LBJL – ibid., Memo. Reedy for Johnson, 5 November 1957. It is not clear whether Furnas was ever offered the post, nor, if so, on what grounds he declined it.
53. Schoettle, 'The Establishment of NASA' (n. 43), p. 233; McDougall, *Heavens* (n. 12), p. 152; R. Evans and R. Novak, *Lyndon B. Johnson: the Exercise of Power* (New York: New American Library, 1966), p. 194; C. C. Alexander, *Holding the Line: the Eisenhower Era 1952–1961* (Bloomington, IN: Indiana U. Press, 1975), p. 217. (Alexander attributes this last description to Evans and Novak, *Johnson*, but his citation is inaccurate and may perhaps refer to another source.)
54. E. M. Emme, *History of Space Flight* (1965 – ch. 5, n. 34), p. 127; Alexander, 'Research' (n. 19), p. 77 – unfortunately, the footnote referred to at this point is misplaced. The most bizarre examples of the tendency of space historians to be misled by Johnson's mastery of verbal nuances both relate to his account of how he spent the evening of 4 October 1957. Simply by saying that he went for a walk, looked up at the stars and *thought* about Sputnik 1, Johnson was able to plant in the minds of two of America's shrewdest space historians the impression that he was claiming to have *seen* the satellite that evening, which would have made him the first person in the United States to do so by about 36 hours: NHO – W. G. Stroud, OH Interview with E. M. Emme, 21 September 1973, pp. 36–7, and J. M. Logsdon, *The Decision to Go to the Moon* (1971 – ch. 11, n. 110), p. 21; c.f. LBJL – Johnson interview with W. Cronkite for CBS, 5 July 1969, p. 2, and L. B. Johnson, *The Vantage Point* (1972 – ch. 2, n. 26), p. 272.
55. Evans and Novak, *Johnson* (n. 53), p. 192.
56. Schoettle, 'Establishment' (n. 43), pp. 228, 221.
57. Logsdon, *Decision* (n. 54), p. 21.
58. McDougall, *Heavens* (n. 12), p. 153.

## Chapter 13: Versions of Events

1. D. D. Eisenhower, *Waging Peace* (1966 – ch. 10, n. 41), pp. 205–26. A private letter by Eisenhower, written in 1965, confirms several of the points made in this section and elsewhere in this chapter.

2. R. M. Bissell, COHC Interview with E. Edwin, 5 June 1967, p. 38; NHO – Grosse Biog. File, Ltr, Grosse to Emme, 12 January 1973.
3. Eisenhower, *Waging Peace* (n. 1), p. 206.
4. S. Alsop, *The Center* (1968 – ch. 10, n. 32), pp. 243–4.
5. Eisenhower, *Waging Peace* (n. 1), pp. 207–8, 212.
6. Ibid., p. 209.
7. Ibid., pp. 209–10.
8. Ibid., p. 209.
9. Ibid., p. 210.
10. J. S. Rigden, *Rabi: Scientist and Citizen* (1987 – ch. 9, n. 9), p. 238.
11. E. N. Hayes, 'Tracking Sputnik 1' in A. C. Clarke (ed.), *The Coming of the Space Age: Famous Accounts of Man's Probing of the Universe* (London: Gollancz, 1967).
12. H. L. Richter, W. F. Sampson and R. Stevens, 'Microlock' (1958 – ch. 8, n. 47); W. H. Pickering with J. H. Wilson, 'Countdown to Space Exploration' (1977 – ch. 8, n. 47); US Congress, Senate, Space Sciences, Hearings: *NASA Authorization for Fiscal Year 1960* pt 1 (1959 – ch. 8, n. 47), testimony of J. T. Mengel, Head of Space Tracking Systems Branch, Vanguard Division, NASA, p. 310; D. E. B. Wilkins, 'From HF Radio to Unified S-Band: an Historical Review of the Development of Communications in the Space Age', paper to 38th IAF Congress, Brighton, October 1987, p. 5.
13. W. S. Bainbridge, *The Spaceflight Revolution* (1976 – ch. 9, n. 92), p. 12.
14. F. I. Ordway and M. R. Sharpe, *The Rocket Team* (1979 – ch. 3, n. 38), p. 362. A recent memoir paints a vivid picture of the extent to which the 'Germans' were disregarded by American engineers carrying out the V-2 launchings of Project Hermes: D. Davis, 'The Coming of Age of U.S. Rocketry', *J. Brit. Interplanetary Soc.*, vol. 39, 1986.
15. F. I. Ordway and F. C. Durant: interviews.
16. W. von Braun and F. I. Ordway, *History of Rocketry & Space Exploration* (1967 – ch. 1, n. 45), pp. 180, 159. Both statements were repeated in the fourth edition: W. von Braun, F. I. Ordway and D. Dooling, *Space Travel* (1985 – ch. 1, n. 46).
17. F. C. Durant: interview.
18. J. M. Logsdon, *The Decision to Go to the Moon* (1970 – ch. 11, n. 110), pp. 153–4.
19. C. C. Furnas, 'Birthpangs of the First Satellite', *Research Trends*, vol. 18, Spring 1970; S. F. Singer: interview.
20. F. L. Whipple, J. A. Van Allen and S. F. Singer: interviews.
21. This statement is not invalidated by the books by Stehling or Van Allen, since neither was in a position to advise the administration on space policy before 1958.
22. C. Mc. Green and M. Lomask, *Vanguard* (1971 – ch. 6, n. 42), pp. 252–3.
23. C. C. Alexander, 'Part One – Research' (1966 – ch. 6, n. 55), p. 29; E.

A. Muenger, *Searching the Horizon: a History of the Ames Research Center 1940–1976* (Washington: NASA, 1985), p. 84.

24.   E. C. B. Schoettle, 'The Establishment of NASA' (1966 – ch. 7, n. 46), pp. 164–5.

25.   J. M. McLeod and J. W. Swinehert, *Satellites, Science, and the Public: Report of a National Survey on the Public Impact of Early Satellite Launchings, directed by Robert C. Davis* (Ann Arbor, MI: U. Michigan Survey Research Center, 1959), p. 1.

26.   Logsdon, *Decision* (n. 18), p. 15.

27.   J. M. Logsdon, 'Opportunities for Policy Historians: the Evolution of the U.S. Civilian Space Program' in A. F. Roland (ed.), *A Spacefaring People* (Washington: NASA, 1985), pp. 83, 82.

28.   Logsdon, ibid., p. 83 – emphasis added.

29.   W. A. McDougall,  . . . *the Heavens and the Earth* (1985 – ch. 1, n. 47).

30.   Ibid., pp. 123–4.

31.   DDEL – WHO: NSC Staff, OCB Central Files, Box 11, OCB 000.91, Folder: 1 (3), Agenda of Meeting for Representatives of Government Agencies in the Satellite Program; US National Security Council, 'U.S. Scientific Satellite Program' (1955 – ch. 7, n. 29), p. 1 par. 2; G. W. Hoover: interview.

32.   DDEL – WHO: NSC Staff, OCB Central Files, Box 11, OCB 000.91, Folder: 1 (3), 'Memorandum of Meeting etc., August 3, 1955', 3 August 1955, p. 2. As noted in Chapter 11, McDougall does not mention the existence of the OCB Working Group before September 1958: *Heavens* (n. 29), pp. 183–4.

33.   For the pre-sputnik discussion of the legal issues, see US Congress, Senate, Space Sciences, Symposium: *Legal Problems of Space Exploration* (87th Congress, 1st Session, 1961), pp. 1–181.

34.   US Congress, ibid., E. Korovin, 'International Status of Cosmic Space', p. 1065; ibid., G. A. Osnitskaya, 'International Law Problems of the Conquest of Space', p. 1090.

35.   US Congress, ibid., G. Zhukov, 'Space Espionage Plans and International Law', p. 1100.

36.   It must be added that before October 1957 the administration was largely in the hands of the press and other self-appointed publicists, many of whom either shared, or made their living by catering to, an uncomplicated national self-esteem with respect to advanced technology, and who therefore preferred to ignore even such information about Soviet progress as they were given. (After the sputniks an element of professional embarrassment may well have contributed to resentment of the administration on the part of the authors of works with titles such as *Vanguard! the Story of the First Man-Made Satellite* or 'The U.S. Earth Satellite Program – Vanguard of Outer Space'.)

37.   *New York Times*, 20 October 1957; c.f. McDougall, *Heavens* (n. 29), p. 148. On 26 January 1956 Richard Porter opened his introductory

presentation to the Ann Arbor conference on 'Scientific Uses of Earth Satellites' with the phrase 'Welcome to the "basketball game"!'.

38. R. S. Lewis, *From Vinland to Mars: a Thousand Years of Exploration* (New York: Quadrangle, 1976), p. 131.

39. A. Wilson, *The Eagle Has Wings: the Story of American Space Exploration 1945–1975* (London: British Interplanetary Society, 1982), p. 9.

40. US National Security Council, 'Scientific Satellite Program' (n. 31), p. 11.

41. T. Osman, *Space History* (London: Michael Joseph, 1985), p. 38.

# Index

There are no general entries for 'Soviet Union' or 'United States' since both are referred to constantly. Institutions, committees etc. are assumed to have been constituted in the United States unless otherwise designated, and are usually listed separately and without the prefix unless it is customary, as in 'United States Information Agency'. For Joint Committees of Congress see 'United States Congress', but for others see 'Senate' or 'House of Representatives'. Congressional subcommittees have been omitted except for the Preparedness Investigating Subcommittee, under 'Senate, Armed Services Committee'. Discursive passages in the Notes are indexed in the same way as the main text, including any commentary on authors and books. First names and other initials are given for individuals wherever possible but all ranks and titles have been omitted.